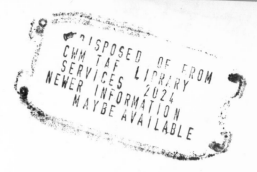
Structure and function
of the body

is due for return on or before the last date

2 3 APR 2007

Structure and function of the body

CATHERINE PARKER ANTHONY, R.N., B.A., M.S.

Formerly Assistant Professor of Nursing, Science Department, and Assistant Instructor of Anatomy and Physiology, Frances Payne Bolton School of Nursing, Case Western Reserve University, Cleveland, Ohio; formerly Instructor of Anatomy and Physiology, Lutheran Hospital and St. Luke's Hospital, Cleveland, Ohio

GARY A. THIBODEAU, Ph.D.

Associate Professor, South Dakota State University, Brookings, South Dakota

SIXTH EDITION

with 186 illustrations

THE C. V. MOSBY COMPANY   ST. LOUIS • TORONTO • LONDON   1980

SIXTH EDITION

Copyright © 1980 by The C. V. Mosby Company

All rights reserved. No part of this book may be reproduced
in any manner without written permission of the publisher.

Previous editions copyrighted 1960, 1964, 1968, 1972, 1976

Printed in the United States of America

The C. V. Mosby Company
11830 Westline Industrial Drive, St. Louis, Missouri 63141

Library of Congress Cataloging in Publication Data

Anthony, Catherine Parker, 1907-
   Structure and function of the body.

   Bibliography: p.
   Includes index.
   1. Human physiology.   2. Anatomy, Human.
I. Thibodeau, Gary A., 1938-   joint author.
II. Title.   [DNLM: 1. Anatomy—Nursing texts.
2. Physiology—Nursing texts.   QS4.3 A628s]
QP34.5.A58   1980   612   79-24006
ISBN 0-8016-0287-4 (Paperback)   03/C/315
ISBN 0-8016-0273-4 (Hardbound)   03/C/336

C/VH/VH 9 8 7 6 5 4 3 2 1   02/A/238

Preface

Structure and Function of the Body is an introductory textbook about the human body. The selection and organization of its content and especially its writing style make this book particularly well suited for courses for practical nurses and other allied health students.

To evaluate this sixth edition you might begin by examining the table of contents. A quick glance will show you that the book is organized not only into chapters but also into groups of chapters or units. Each unit develops one major theme, a theme clearly identified by the unit's name. The content of the book as a whole is organized around the seven major themes designated by the units. The content of every chapter is also clearly organized. Brief introductory outlines alert students to the main topics discussed in the chapter, and text headings call attention to the chapter's text, summarizing it and emphasizing and tying together its contents. Finally, key questions follow the summarizing outline to help students with those difficult tasks of identifying and learning essential information. How much does a textbook's organization contribute to its effectiveness? Immeasurably!

In our view, students take their first giant step toward understanding and learning when they perceive clearly the organization of material.

The organization of this sixth edition differs somewhat from the previous one. The main changes are the following: a new unit titled "Systems that Provide Transportation and Immunity"; separate chapters about the blood, the cardiovascular and lymphatic systems, and the somatic and autonomic divisions of the nervous system; and a new chapter on the immune system.

Many changes in content appear in virtually every chapter of this edition. A few examples of material added are discussions of diffusion, osmosis, filtration, active transport, mitosis, neurotransmitters, prostaglandins, cardiopulmonary resuscitation, and clinical applications.

The many new color illustrations created for this edition by the eminent medical artist, Ernest W. Beck, have added considerable beauty to the book. Not only will they make it easier for students to understand and learn much of the text, but they will also surely make them enjoy it more.

Some final notes about this edition—unlike previous editions it contains a list of abbreviations, prefixes, and suffixes commonly used in medical literature. Also, for the first time instructors will be supplied a test manual.

As senior author, I acknowledge with warm appreciation the valuable contributions made by my new coauthor, Dr. Gary A. Thibodeau, by the artist, Mr. Beck, and by the hosts of people who have produced this book and who will distribute it.

Catherine Parker Anthony

Contents

Structure and function of the body

The body as a whole

chapter 1

An introduction to the structure and function of the body

Wonders are many in our world today, but none is more wondrous than the human body. This is a textbook about that incomparable structure. Chapter 1 relates a few basic facts about the body's structure and its functions. It also defines some of the words scientists use in talking about the body.

Some basic facts about body structure

1 The body is a single structure, but it is made up of billions of smaller structures of four major kinds: cells, tissues, organs, and systems. The smallest and most numerous of these are cells. Although long recognized as the simplest units of living matter, cells are, in truth, far from simple. They are extremely complex, a fact you will discover in Chapter 2.

Tissues are somewhat more complex units than cells. By definition, a tissue is an organization of a great many similar cells with varying amounts and kinds of nonliving, intercellular substances between them.

Organs are more complex units than tissues. An organ is a group of several different kinds of tissues so arranged that together they can perform a special function. For example, the stomach is an organization of muscle tissue, connective tissue, epithelial tissue, and nervous tissue so arranged that together they can perform part of the function of digestion.

Systems are the most complex of the units that make up the body. A system is an organization of varying numbers and kinds of organs so arranged that together they can perform complex functions for the body. For example, the digestive system consists of the mouth, esophagus, stomach, intestines, and a few other structures so arranged that together they can perform the complex functions of digestion and absorption.

2 Organization is a most important characteristic of body structure. No haphazard conglomerations, but only orderly arrangements, exist in the body and in every living thing. Even the name "organism," used to denote a living thing, implies organization.

3 The body, contrary to its external appearance, is not a solid structure. It contains four cavities, which in turn contain compact, well-ordered arrangements of internal organs. To discover the names and locations of the cavities, look at Fig. 1-1. What anatomical name do you find for the space that you perhaps think of as your chest cavity? The midportion of this cavity is named the *mediastinum* and its other divisions are the right and the left *pleural cavities*. Fig. 1-1 also identifies an abdominal cavity and a pelvic cavity. Actually they form only one cavity—the *abdominopelvic cavity*—since no partition of any kind separates them. Notice, however, that a partition does separate the thoracic cavity from the abdominal cavity. It is a dome-shaped muscle and is the most important muscle we have for breathing. What is its name? The space inside the skull contains the brain and is called the *cranial cavity*. The space inside the spinal column is called the *spinal cavity;* it contains the spinal cord. The cranial and spinal cavities are classified as *dorsal cavities,* whereas the thoracic and abdominopelvic cavities are called *ventral cavities.* Dorsal (or posterior in humans) means "back." Ventral (or anterior in humans) means "front," that is, the stomach side.

Fig. 1-2 shows some of the organs contained in the largest body cavities. For example, it shows the trachea, aorta, and heart in the mediastinal portion of the thoracic cavity and the lungs in the pleural portions. Observe the many organs shown in the abdominal cavity: the liver, gallbladder, stomach, spleen, pancreas, and parts of the small intestine (cecum and ascending, transverse, and descending colon). The sigmoid colon and rectum lie in the pelvic cavity. Among the organs not visible in Fig. 1-2 are the thymus gland in the mediastinum, the urinary bladder, certain reproductive organs in the pelvic cavity, and numerous blood vessels and nerves.

Find each body cavity in a model of the human body if you have access to one. Try

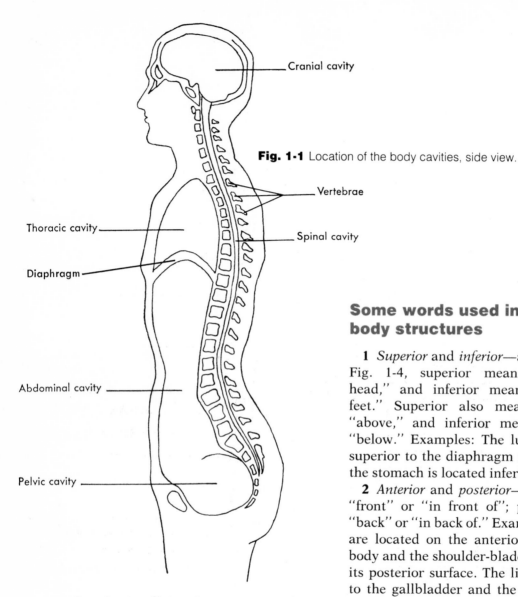

Cranial cavity

Fig. 1-1 Location of the body cavities, side view.

Vertebrae

Thoracic cavity

Spinal cavity

Diaphragm

Abdominal cavity

Pelvic cavity

Some words used in describing body structures

1 *Superior* and *inferior*—as you can see in Fig. 1-4, superior means "toward the head," and inferior means "toward the feet." Superior also means "upper" or "above," and inferior means "lower" or "below." Examples: The lungs are located superior to the diaphragm muscle, whereas the stomach is located inferior to it.

2 *Anterior* and *posterior*—anterior means "front" or "in front of"; posterior means "back" or "in back of." Examples: The knees are located on the anterior surface of the body and the shoulder-blades are located on its posterior surface. The liver lies anterior to the gallbladder and the gallbladder lies posterior to the liver. The terms anterior and posterior are used in referring to a human. In reference to animals, ventral and dorsal are substituted for anterior and posterior in man.

3 *Medial* and *lateral* (Fig. 1-3)—medial means "toward the midline of the body"; lateral means "toward the side of the body or away from its midline." Examples: The great toe is located at the medial side of the foot and the little toe is located at its lateral

to identify the organs in each cavity, and try to visualize their locations in your own body. Study Figs. 1-1 and 1-2.

4 The structure of the body changes in many ways and at varying rates during a lifetime. Before young adulthood it develops and grows. After young adulthood it gradually undergoes various changes and, in general, atrophies. Nearly every chapter of this book will mention a few of these changes.

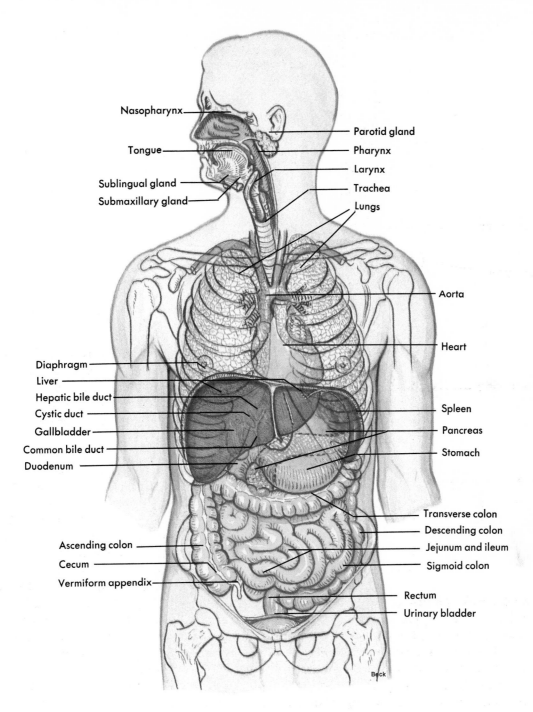

Fig. 1-2 Organs of the thoracic and abdominal cavities viewed from the front.

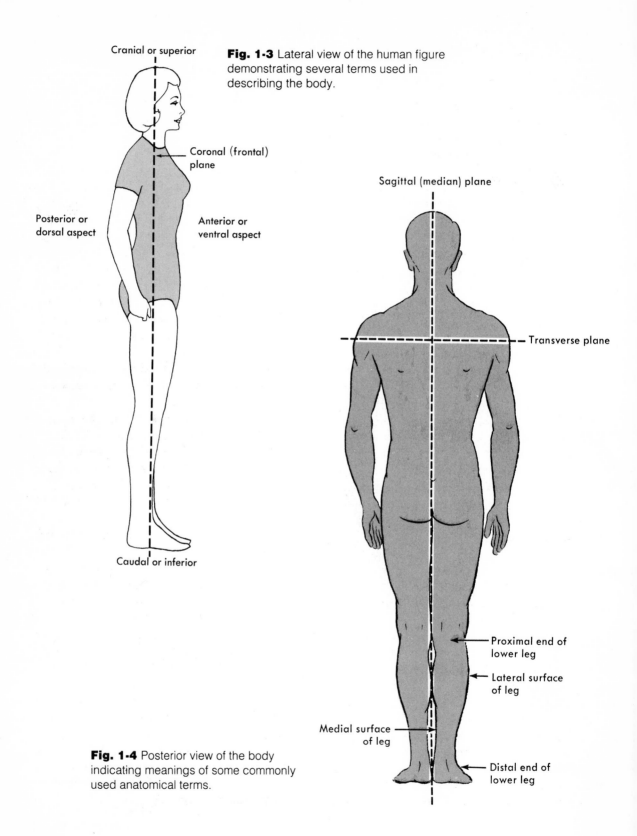

Cranial or superior

Fig. 1-3 Lateral view of the human figure demonstrating several terms used in describing the body.

Coronal (frontal) plane

Posterior or dorsal aspect

Anterior or ventral aspect

Caudal or inferior

Fig. 1-4 Posterior view of the body indicating meanings of some commonly used anatomical terms.

Sagittal (median) plane

Transverse plane

Proximal end of lower leg

Lateral surface of leg

Medial surface of leg

Distal end of lower leg

side. The heart lies medial to the lungs and the lungs lie lateral to the heart.

4 *Proximal* and *distal*—proximal means "toward or nearest the trunk of the body, or nearest the point of origin of one of its parts"; distal means "away from or farthest from the trunk or the point of origin of a body part." Examples: The elbow lies at the proximal end of the lower arm, whereas the hand lies at its distal end.

5 *Sagittal,* *frontal,* and *transverse planes*—a sagittal plane is a lengthwise plane running from front to back; a cut made through the sagittal plane of an organ therefore divides it into right and left sections. A frontal plane is a lengthwise plane running from side to side as indicated in Fig. 1-4. Cutting an organ through its frontal plane divides it into anterior and posterior sections. A transverse plane is a horizontal or crosswise plane. Cutting an organ through the transverse plane divides it into upper and lower sections.

6 *Abdominal regions*—to make it easier to locate abdominal organs, anatomists have divided the abdomen into the nine regions shown in Fig. 1-3 and defined them as follows:

  a *Upper abdominal regions*—right hypochondriac, epigastric, and left hypochondriac regions; lie above an imaginary line across the abdomen at the level of the ninth rib cartilages

  b *Middle regions*—right lumbar, umbilical, and left lumbar regions; lie below an imaginary line across the abdomen at the level of the ninth rib cartilages and above an imaginary line across the abdomen at the top of the hipbones

  c *Lower regions*—right iliac, hypogastric, and left iliac regions; lie below an imaginary line across the abdomen at the level of the top of the hipbones (see Fig. 1-5)

7 *Anatomical position*—if you stand erect with your arms at your sides and the palms of your hands turned forward, your body is in the anatomical position.

Some basic facts about body functions

1 Survival is the body's most important business—survival of itself and survival of the human species. Although the body carries on a great many different functions, each one contributes in some way to survival of the individual or of humankind.

2 Survival depends on the body's maintaining or restoring homeostasis. Homeostasis means relative constancy of the internal environment. More specifically, homeostasis means that the chemical composition, the volume, and certain other characteristics of blood and interstitial fluid (fluid around cells) remain constant within narrow limits.

3 Homeostasis depends on the body's ceaselessly carrying on many activities. It must continually respond to changes in its environment, exchange materials between its environment and its cells, metabolize food, and control all of its diverse activities.

4 All body functions are ultimately cell functions.

5 Body functions are related to age. During childhood, body functions gradually become more and more efficient and effective. They operate with maximum efficiency and effectiveness during young adulthood. During late adulthood and old age, they gradually become less and less efficient and effective. Changes and functions that occur during the early years are called developmental processes. Those that occur after young adulthood are called aging processes. In general, developmental processes improve functions. Aging processes usually diminish them. Future chapters will refer to many functional age changes.

Fig. 1-5 The nine regions of the abdomen. The top horizontal line crosses the abdomen at the level of the ninth rib cartilages. The lower horizontal line crosses it at the level of the iliac crests. The vertical lines pass through the midpoints of the right and left inguinal (Poupart's) ligaments.

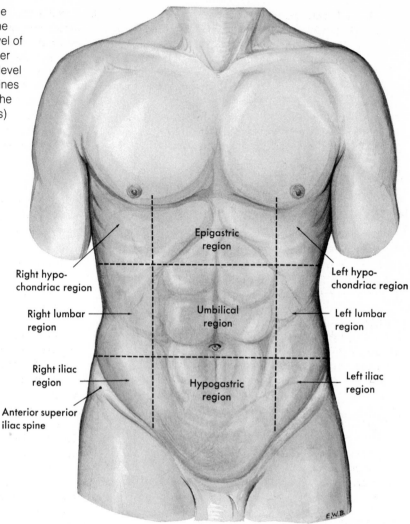

Right hypo-chondriac region

Right lumbar region

Right iliac region

Anterior superior iliac spine

Epigastric region

Umbilical region

Hypogastric region

Left hypo-chondriac region

Left lumbar region

Left iliac region

Outline summary

Some basic facts about body structure

A The body is a unit constructed of the following smaller units—cells, tissues, organs, systems
  1 Cells—the smallest structural units; organizations of various chemicals
  2 Tissues—organizations of similar cells
  3 Organs—organizations of different kinds of tissues
  4 Systems—organizations of different kinds of organs

B Organization is an outstanding characteristic of body structure
C The presence of the following cavities is a prominent feature of body structure
  1 Anterior cavities
    a Thoracic cavity
      (1) Mediastinum—midportion of thoracic cavity; heart, trachea, esophagus, and thymus gland located in mediastinum
      (2) Pleural cavities—right lung located in right pleural cavity, left lung in left pleural cavity

    b Abdominopelvic cavity
      (1) Abdominal cavity contains stomach, intestines, liver, gallbladder, pancreas, and spleen
      (2) Pelvic cavity contains reproductive organs, urinary bladder, lowest part of intestine
  2 Posterior cavities
    a Cranial cavity contains brain
    b Spinal cavity contains spinal cord

Some words used in describing body structures

A Superior—toward the head, upper, above
Inferior—toward the feet, lower, below
B Anterior—front, in front of
Posterior—back, in back of
C Medial—toward midline of a structure
Lateral—away from midline or toward side of a structure
D Proximal—toward or nearest trunk, or nearest point or origin of a structure
Distal—away from or farthest from trunk, or farthest from a structure's point of origin
E Sagittal plane—lengthwise plane that divides a structure into right and left sections
Frontal plane—lengthwise plane that divides a structure into anterior and posterior sections
Transverse plane—horizontal plane that divides a structure into upper and lower sections
F Abdominal regions—see Fig. 1-5
G Anatomical position—standing erect with arms at sides and palms turned forward

Some basic facts about body functions

A Survival is the body's most important business—survival of the individual and of the human species
B Survival depends on the maintenance or restoration of homeostasis (relative constancy of the internal environment)
C Homeostasis depends on never-ceasing activities: response to changes in the external and internal environment, exchange of materials between the environment and cells, metabolism, and control
D All body functions are ultimately cell functions
E Body functions are related to age; peak efficiency during young adulthood, diminishing efficiency after young adulthood

Review questions

1 Name the four kinds of structural units of the body. Define each briefly.
2 In what cavity could you find each of the following?

appendix	liver
brain	lungs
esophagus	pancreas
gallbladder	spinal cord
heart	spleen
intestines	urinary bladder

3 In one word, what is the one dominant function of the body or of any living thing?
4 Explain briefly what the term "homeostasis" means.
5 What do the terms "proximal" and "distal" mean?
6 Besides the maintenance of homeostasis and the carrying on of metabolism, name another major function the body must perform in order to survive.
7 On what surface of the body are the toenails located?
8 What structures lie lateral to the bridge of the nose?
9 Which joint—hip or knee—lies at the distal end of the thigh?
10 Define the term "anatomical position."

chapter 2

Cells and tissues

About 300 years ago Robert Hooke looked through his microscope—one of the very early, somewhat primitive ones—at some plant material. What he saw must have surprised him. Instead of a single magnified piece of plant material, he saw many small pieces. Since they reminded him of miniature prison cells, that is what he called them—cells. Since Hooke's time, thousands of individuals have examined thousands of plant and animal specimens and found them all, without exception, to be composed of cells. This fact, that cells are the smallest structural units of living things, has become the foundation stone of modern biology. Many living things are so simple that they consist of just one cell. The human body, however, is so complex that it consists not of a few thousands, or millions, or even billions of cells, but of many trillions of them. This chapter discusses cells first and then tissues.

Cells

Composition

Cells are composed of protoplasm or "living matter," a substance that exists only in cells. It consists mostly of water. In the water are four main compounds found only in protoplasm. They are proteins, carbohydrates, fats (lipids), and nucleic acids. Protoplasm also contains small amounts of ordinary table salt, several other salts, and a few other kinds of compounds.

Size and shape

Human cells are microscopic in size; that is, they can be seen only when magnified by a microscope. However, they vary considerably in size. An ovum (female sex cell), for example, has a diameter of a little less than 1,000 micrometers* (about $^1/_{25}$ of an inch), whereas red blood cells have a diameter of only 7.5 micrometers. Cells differ even more markedly in shape than in size. Some are flat, some are brick-shaped, some are threadlike, and some have irregular shapes.

Structural parts

The three main parts of a cell are the cytoplasmic membrane, cytoplasm, and nucleus. The *cytoplasmic membrane,* as its name suggests, is the membrane that encloses the cytoplasm. It forms the outer boundary of the cell. (Plasma membrane is another name for cytoplasmic membrane.) It is an incredibly delicate structure—only

*A micrometer is one millionth of a meter. (Micron is another name for micrometer.) In the metric system the units of length are as follows:

1 meter (m) = 39.37 inches
1 centimeter (cm) = 1/100 meter
1 millimeter (mm) = 1/1,000 meter
1 micrometer (μm) or micron (μ) = 1/1,000,000 meter
1 nanometer (nm) = 1/1,000,000,000 meter
1 Angstrom (Å) = 1/10,000,000,000 meter
Approximately equal to 1 inch:
2.5 cm
25 mm
25,000 μm
25,000,000 nm
250,000,000 Å

about 3/10,000,000 of an inch thick! And yet it has a precise, orderly structure. According to a widely held concept, two layers of phosphate-containing fat molecules (phospholipids) form a fluid framework for the cytoplasmic membrane. Protein molecules lie at both outer and inner surfaces of this framework, and many extend all the way through it. Despite its seeming fragility, the cytoplasmic membrane is strong enough to keep the cell whole and intact. It also performs other life-preserving functions for the cell. It serves as a well-guarded gateway between the fluid inside the cell and the fluid around it. It allows certain substances to move through it but bars the passage of others. A recent discovery is that the cytoplasmic membrane even functions as a communication device. How? Some of the protein molecules on the membrane's outer surface serve as receptors for certain other molecules when they contact them. In other words, certain molecules bind to certain receptor proteins. For example, some hormones (chemicals secreted into blood from ductless glands) bind to membrane receptors and a change in cell functions follows. We might therefore think of such hormones as chemical messages, communicated to cells by binding to their cytoplasmic membrane receptors.

Cytoplasm is the protoplasm that lies between the cytoplasmic membrane and the nucleus (the sphere-shaped structure in the center of the cell). Numerous small structures are located in the cytoplasm. As a

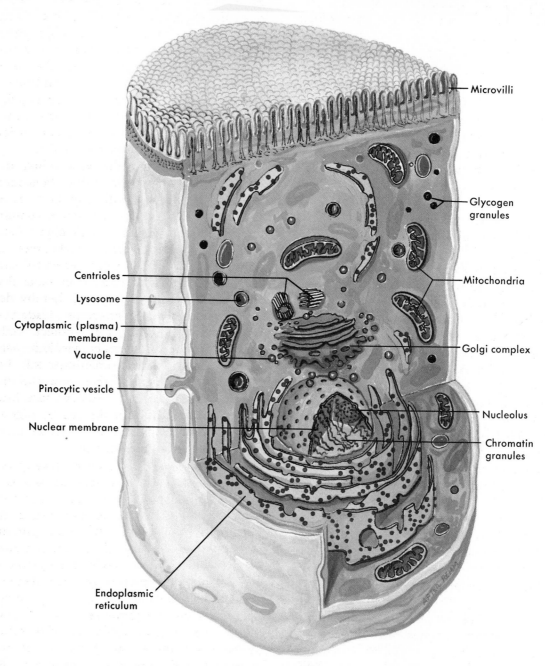

Centrioles
Lysosome
Cytoplasmic (plasma) membrane
Vacuole
Pinocytic vesicle
Nuclear membrane
Endoplasmic reticulum

Microvilli
Glycogen granules
Mitochondria
Golgi complex
Nucleolus
Chromatin granules

Fig. 2-1 Artist's interpretation of cell structure as seen under an electron microscope. Note the many mitochondria, popularly known as the "power plants of the cell." Note, too, the innumerable dots bordering the endoplasmic reticulum. These are ribosomes, the cell's "protein factories."

group they are called *organelles,* that is, little organs, an appropriate name since they function for the cell much as organs function for the body.

Look now at Fig. 2-1. Notice how many different kinds of structures you can see in the cytoplasm of this cell. A generation ago almost all of these organelles were unknown. They are so small that even when magnified 1,000 times by a light microscope, they are invisible. Electron microscopes brought them into view by magnifying them many thousands of times. Perhaps you are already familiar with the names of the organelles we shall describe: endoplasmic reticulum, ribosomes, mitochondria, lysosomes, and Golgi apparatus.

The *endoplasmic reticulum* is a network of connecting sacs and canals that wind tortuously through a cell's cytoplasm, all the way from its cytoplasmic membrane to its nucleus. Fig. 2-1 shows only a small part of the endoplasmic reticulum. This organelle functions as a miniature circulatory system for the cell.

Did you notice the dots scattered along the walls of the endoplasmic reticulum in Fig. 2-1? These small but vitally important organelles are called *ribosomes.* Every cell contains thousands of them. Many of them lie free in the cytoplasm unattached to the endoplasmic reticulum. Ribosomes carry on a most complex function, that of making enzymes and other protein compounds. Their nickname—"protein factories"—indicates this function.

Mitochondria are another kind of organelle present in all cells. Observe their appearance in Fig. 2-1, where they are magnified thousands of times. Mitochondria are so tiny that a lineup of 15,000 or more of them would fill a space only 1 inch long. Two membranous sacs, one inside the other, compose a mitochondrion. The inner membrane forms folds that look like miniature incomplete partitions. Within a mitochondrion's fragile walls, complex, energy-releasing chemical reactions go on continuously. Because these reactions supply most of the power for doing cellular work, mitochondria have been nicknamed the cell's "power plants." Survival of cells and therefore of the body depends on mitochondrial chemical reactions. Without these reactions, cells would soon have no energy for doing the work that keeps them alive.

Lysosomes are membranous-walled organelles, which in their active stage look like small sacs, often with tiny particles in them (Fig. 2-1). Because lysosomes contain chemicals (enzymes) that can digest food compounds, one of their nicknames is "digestive bags." Lysosomal enzymes can also digest substances other than foods. For example, they can digest and thereby destroy microbes that manage to invade the cell. Thus lysosomes can protect cells against destruction by microbes. But, paradoxically, lysosomes sometimes kill instead of protect cells. If their powerful enzymes escape from the lysosome sacs into the cytoplasm, they kill the cell by digesting it. This fact has earned lysosomes their other nickname—"suicide bags."

The *Golgi apparatus,* long a mystery organelle, consists of tiny sacs stacked one upon the other near the nucleus. It is now known to make certain carbohydrate compounds, combine them with certain protein molecules, and package the product in neat little globules. Then slowly these globules move outward to and through the cell membrane. Once outside the cell they break open, and their contents spill out. An example of a Golgi apparatus product is the slippery substance called mucus. If we wanted to nickname the Golgi apparatus, we might call it the cell's "carbohydrate-producing and -packaging factory."

The *nucleus* of a cell, viewed under a

light microscope, looks like a very simple structure indeed—just a small sphere in the central portion of the cell. In some cases dark granules or rodlike structures can be seen inside the nucleus. These are the cell's chromosomes, and their small size greatly belies the importance of their function, heredity. (See pp. 15-18.)

Functions

Every human cell performs certain functions that maintain its own survival and others that help maintain the body's survival. Many, but not all, cells perform another kind of function that maintains survival of their species—they reproduce themselves. Of the functions that maintain a cell's own life, we shall discuss in this chapter only the major processes by which various substances move through the cell membrane, namely, diffusion, osmosis, filtration, and active transport. Then we shall name a few examples of functions cells contribute to survival of the body. Finally, we shall describe cell reproduction.

Movement of substances through cell membranes

DIFFUSION. Diffusion is the process by which dissolved particles (solutes) scatter themselves evenly throughout a fluid. Also, diffusion is the process by which solutes and water move through a membrane to distribute themselves evenly throughout the two fluids separated by the membrane. To demonstrate that solute particles diffuse through a fluid, perform this simple experiment the next time you pour yourself a cup of coffee or tea. Place a cube of sugar on a teaspoon and lower it gently to the bottom of the cup. Let it stand 2 or 3 minutes, and then, holding the cup level, take a sip off the top. It will taste sweet. Why? Because some of the sugar molecules have diffused up to the top of the beverage away from the bottom where they were originally concen-

trated. Given enough time, diffusion will scatter the sugar evenly throughout the fluid.

Diffusion through a membrane scatters solutes evenly throughout the fluids separated by the membrane. For example, suppose a membrane permeable to both salt and water separates a 20% salt solution from a 10% salt solution. (A membrane that is permeable to a substance allows that substance to pass through it.) Salt particles and water molecules race around in all directions in both salt solutions. Inevitably they bump into each other and into the membrane on both sides. Because the 20% solution contains more salt particles, more of them hit the 20% side of the membrane than the 10% side and therefore more salt particles diffuse out of the 20% solution into the 10% solution than diffuse in the opposite direction. The 20% solution loses salt and the 10% solution gains it, and eventually the concentrations of the two solutions become equal. What should you remember about diffusion? The following facts:

1 Diffusion is the movement of both solute particles and water molecules through a fluid or a membrane.

2 Diffusion results eventually in an even distribution of solute particles in a fluid or in two fluids separated by a membrane. Diffusion across a membrane eventually results in equilibration of the solutions separated by the membrane. In other words, diffusion makes their solute concentrations become equal.

OSMOSIS. Osmosis is the diffusion of water through a selectively permeable membrane. A *selectively permeable membrane* is one through which some solutes diffuse freely and others do not. The membrane seems to select which solutes can pass through it. Solutes that diffuse freely through a membrane, as we have seen, equilibrate across the membrane; their concentrations on both sides of the membrane become equal. Solutes that do not diffuse freely through a membrane, on the other hand, do not equilibrate across the membrane but maintain a concentration gradient (difference) across it. A higher concentration of a not freely diffusible solute is maintained on one side of a membrane than on the other. Normal living cell membranes are selectively permeable. They maintain certain solute concentration gradients and therefore water moves through them by osmosis.

In which direction does water osmose through selectively permeable membranes? Water osmoses in both directions, but more water osmoses into the fluid with the higher solute concentration than osmoses in the opposite direction. Suppose, for example, that a membrane impermeable to salt separates a 20% salt solution from a 10% salt solution. More water will osmose into the more concentrated 20% solution than will osmose in the opposite direction. A higher solute concentration might be thought of as a magnet that attracts water; the higher the solute concentration, the greater its water-drawing power. What should you remember about osmosis? The following facts:

1 Osmosis is the movement of water through a selectively permeable membrane, a membrane that maintains at least one solute concentration gradient across it.

2 Water osmoses in both directions through a membrane, but more water osmoses into the fluid that originally had the higher solute concentration than osmoses in the opposite direction.

FILTRATION. Filtration is the movement of both water and solutes through a membrane owing to a greater pushing force on one side of the membrane. For example, blood exerts a greater pushing force on the membranous walls of capillaries than tissue (interstitial) fluid exerts on the opposite side of the capillary membrane. Water and solutes therefore filter out of blood in the capillaries into interstitial fluid.

ACTIVE TRANSPORT. Active transport is the movement of a substance through a living cell membrane in an uphill direction. "Uphill" means up a concentration gradient, that is, from a lower to a higher concentration. Because chemical reactions within a cell supply the energy for moving a substance up its concentration gradient, active transport mechanisms can take place only through living membranes. Active transport mechanisms are often called pumps, an appropriate name since it suggests that active transport moves a substance in an uphill direction just as a water pump, for example, moves water. The sodium pump moves sodium through the cytoplasmic membrane from the inside of the cell where sodium concentration is low to the outside of the cell where sodium concentration is high. Would you expect sodium to diffuse through the membrane in this direction? Check your answer by rereading the paragraph on the direction of diffusion on p. 14.

Cell reproduction. Human cells, other than sex cells, reproduce by a process called *mitosis*. In this process a cell divides in order to multiply. One cell divides to form two cells. But before a cell divides, the chromosomes in its nucleus undergo certain changes. Chromosomes are composed largely of a compound named deoxyribonucleic acid but almost always called DNA. DNA is probably the most important compound in the world. We can justify such an extravagant claim by stating DNA's nickname, the heredity molecule. DNA makes heredity possible. Its molecules pass on the capabilities for developing the same characteristics as the parent cells from one generation of cells to the next. By so doing, DNA molecules transmit from each generation of parents to their children all the traits, both physical and mental, that they inherit from their ancestors.

Stating this function is a simple and easy matter, but explaining it is not. It requires the telling of a long, complicated, and still unfinished story. We shall attempt only a brief synopsis.

The main theme of the DNA story revolves around the strange and unique structure of the DNA molecule. To try to visualize the shape of the DNA molecule, picture first an extremely long, narrow ladder made of a pliable material. Now see it twisting round and round on its axis and taking on the shape of the DNA molecule—a double helix (Greek word for spiral).

As to its structure, the DNA molecule is made up of many smaller units, namely, a sugar, bases, and phosphate units. As you can see in Fig. 2-2, the name of the sugar is deoxyribose. The names of the bases are adenine, thymine, guanine, and cytosine. Observe that the sides of the DNA ladder consist of deoxyribose alternating with phosphate units. Look next at the ladder's steps. Notice that each one consists of a pair of bases. Only two combinations of bases occur. The same two bases invariably pair off with each other in a DNA molecule like teenagers "going steady." Adenine and thymine always go together, as do guanine and cytosine. This characteristic of DNA structure is called *complementary base pairing.* It is an important fact that we shall refer to again when we describe DNA's function.

Another important fact about DNA struc-

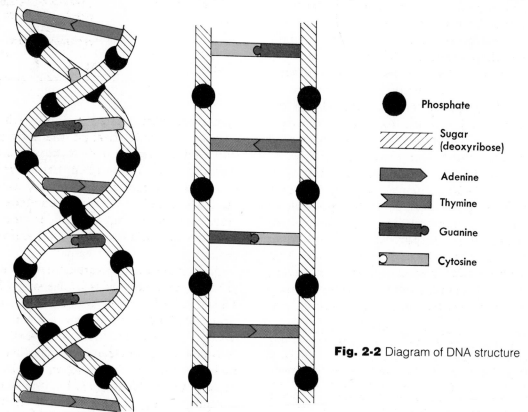

Fig. 2-2 Diagram of DNA structure

ture is that the sequence of its base pairs is not the same in all DNA molecules. The base pairs are the same in all DNA molecules but not their sequence. This fact has tremendous functional importance because it is the sequence of base pairs that determines heredity. In fact that is what a *gene* is—a specific sequence of an average of about 1,000 base pairs. Approximately 100,000 genes, according to one estimate, compose the DNA molecule of just one human chromosome. If we do some simple multiplication, we arrive at some staggering figures:

1 gene, a sequence of 1,000 base pairs.

1 chromosome consists of 100,000 genes.

1 chromosome consists of (100,000 × 1,000) or 100 million base pairs.

1 human cell contains 46 chromosomes.

1 human cell contains (46 × 100,000) or over $4\frac{1}{2}$ million genes.

Is it much wonder, then, with more than $4\frac{1}{2}$ million genes or "heredity-bearers" in each of our cells, that no two of us inherit exactly the same traits?

How do genes bring about heredity? There is, of course, no short and easy answer to that question. In general, genes tell ribosomes* what enzymes and other proteins they are to make. According to a current theory, each gene—or, in other words, each sequence in a DNA molecule of a thousand or so base pairs—is a code indicating the structure of a polypeptide, that is, a part of a protein molecule. This information is relayed to the ribosomes, and these cellular protein factories then make the enzymes and other proteins. En-

*Ribosomes contain a chemical named ribonucleic acid (abbreviation RNA). RNA resembles DNA in that both are composed of four bases, a sugar, and phosphate. RNA, however differs from DNA in three ways: RNA molecules contain the base uracil instead of thymine; RNA molecules contain the sugar ribose instead of deoxyribose; RNA molecules are smaller than DNA molecules.

Table 2-1 Functions of some cell structures

Cell structures	Main functions
Cytoplasmic membrane	Controls entrance and exit of substances into and out of cell; maintains cell's wholeness
Mitochondria	Catabolism
Lysosomes	Digestion
Ribosomes	Anabolism; make protein compounds
Endoplasmic reticulum	Transportation
Golgi apparatus	Makes carbohydrates, combines them with proteins, and packages product
Nucleus	Heredity

zymes are vital substances. Their job is to keep innumerable chemical reactions going on at a fast enough pace to keep cells and therefore the body alive. In summary, genes control enzyme production, enzymes facilitate cellular chemical reactions, and cellular chemical reactions determine both cell structure and function and therefore heredity.

DNA molecules possess a unique ability that no other molecules in the world have. They can make copies of themselves, a process called DNA replication. Before a cell divides to form two new cells, each DNA molecule in its nucleus forms another DNA molecule just like itself. When a DNA molecule is not replicating, it has the shape of a tightly coiled double helix. As it begins the process of replication, short segments of the DNA molecule uncoil and the two strands of the molecule pull apart between their base pairs. The separated strands therefore contain unpaired bases. Each unpaired base in each of the two separated strands of the DNA molecule attracts its complementary base (present in the nuclear fluid) and binds to it. Specifically, each adenine attracts and binds to a thymine and each cytosine attracts and binds to a guanine. These steps repeat themselves over and over again throughout the length of the DNA molecule. Thus each half of a DNA molecule becomes a whole DNA molecule identical to the original DNA molecule.

A cell is ready to reproduce itself (by the process of mitosis) after the DNA molecules in its nucleus have duplicated themselves. DNA replication has transformed each chromosome into two chromatids composed of identical DNA molecules (See Fig. 2-3, *Prophase.*) During mitosis, the two chromatids separate. (See Fig. 2-3, *Anaphase.*) One member of each pair becomes a chromosome in one of the new cells formed when the original cell divides. (See Fig. 2-3, *Telophase.*) The other member of each pair of chromatids becomes a chromosome in the other new cell. Each of the new cells therefore contains the same number of chromosomes as the parent cell. Because the chromosomes are made up of DNA molecules identical to those in the parent cell, each daughter cell has the ability to become like

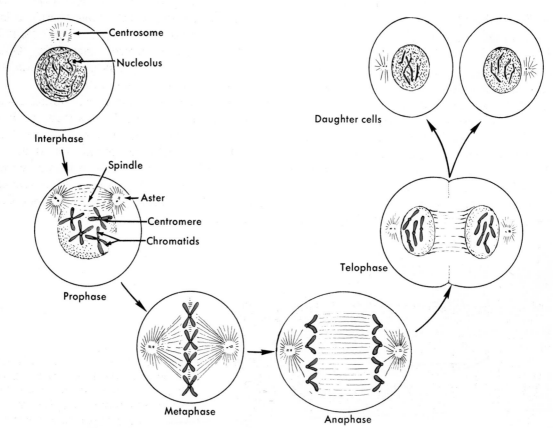

Fig. 2-3 Diagram of mitosis in an animal cell with four chromosomes.

its parent cell. Briefly then, mitosis enables cells to reproduce their own kind.

Tissues

Four main kinds of tissue compose the body's many organs. Their names are epithelial tissue, connective tissue, muscle tissue, and nervous tissue. Tissues differ from each other in the size and shape of their cells, in the amount and kind of material present between the cells, and in the special functions they perform to help maintain the body's survival. We shall name the major locations and types of tissues and describe their structure and functions briefly.

Epithelial tissue

Epithelial tissue covers the body and many of its parts. It also lines various parts of the body. Because epithelial cells are packed close together with little or no intercellular material between them, they form continuous sheets of material that contain no blood vessels. Three of the main types of epithelial tissue are simple squamous epithelium, stratified squamous epithelium, and simple columnar epithelium.

Simple squamous epithelium. Simple squamous epithelium consists of a single layer of scalelike cells. Because of this structure, substances can readily pass through simple squamous epithelial tissue, making absorption its special function. Absorption of oxygen into the blood, for example, takes place through the simple squamous epithelium that forms the tiny air sacs in the lungs.

Stratified squamous epithelium. Stratified squamous epithelium (Fig. 2-4) consists of several layers of closely packed cells, an arrangement that makes this tissue a specialist at protection. For instance, stratified squamous epithelial tissue protects the body against invasion by microorganisms. Most microbes cannot work their way through a barrier of stratified squamous tissue such as that which composes the surface of skin and of mucous membranes. One way of preventing infections therefore is to take good care of your skin. Don't let it become cracked from chapping. Guard against cuts and scratches—particularly good advice for nurses to remember.

Simple columnar epithelium. Simple columnar epithelium lines the stomach and intestine and parts of the respiratory tract. A single layer of cells composes this tissue, but two types of cells—goblet and columnar—may be present (Fig. 2-5). Goblet cells specialize in secreting mucus, whereas columnar cells specialize in absorption.

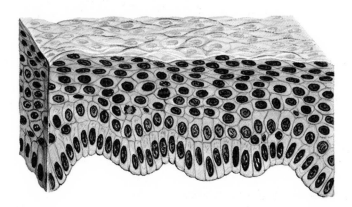

Fig. 2-4 Stratified squamous epithelium. Note the several layers of closely packed cells. Because of these structural features, stratified squamous epithelial tissue protects underlying structures against mechanical injury and microbe invasion.

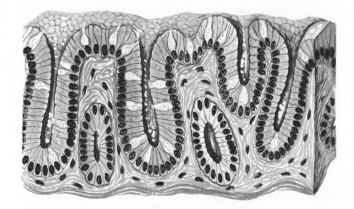

Fig. 2-5 Simple columnar epithelium with goblet cells, such as that which lines intestines. Secretion is the special function of this tissue.

Fig. 2-6 Fat cells. In fat cells, which make up adipose tissue, a fat droplet occupies nearly the entire area of the cell, pushing the cytoplasm and nucleus out to the periphery of the cell.

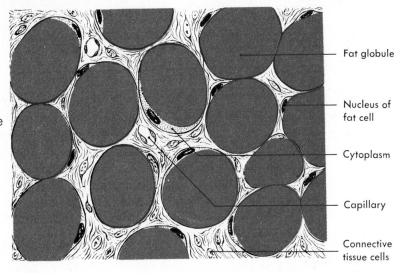

— Fat globule

— Nucleus of fat cell

— Cytoplasm

— Capillary

— Connective tissue cells

Connective tissue

Connective tissue is located in virtually all parts of the body—in the skin, mucous membrane, muscles, bones, nerves, and all internal organs. Among the half dozen or more types of connective tissue are areolar tissue, adipose or fat tissue, dense fibrous tissue, bone, and cartilage.

Connective tissue differs from epithelial tissue in the arrangement and variety of its cells and in the amount and kinds of material between its cells. Connective tissue cells are not packed close together to form continuous sheaths but for the most part lie well separated from each other by intercel-

lular material. In adipose·tissue, for example, a moderate amount of a soft gel-like substance lies between fat cells (Fig. 2-6). Intercellular material is more abundant in areolar tissue and it contains numerous fibers and several kinds of cells (Fig. 2-7). Bone and cartilage are hard tissues because of the nature and amount of their intercellular substance.

Connective tissues, as the name suggests, function to connect various parts of the body. They also provide the supporting framework of the body as a whole and of many of its organs.

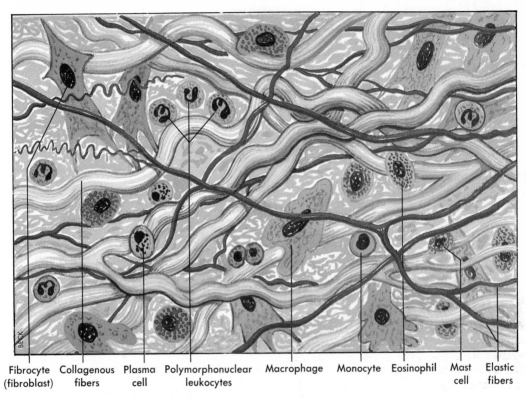

Fibrocyte (fibroblast) Collagenous fibers Plasma cell Polymorphonuclear leukocytes Macrophage Monocyte Eosinophil Mast cell Elastic fibers

Fig. 2-7 Areolar connective tissue. The large white fibers are collagenous fibers. Each of the red strands consists of a bundle of elastic fibers. Several fibroblasts are shown between the fibers. Also shown are macrophages, a plasma cell, a mast cell, and three types of white blood cells: polymorphonuclear leukocytes, eosinophils, and a monocyte.

Muscle tissue

There are three kinds of muscle tissue, namely, skeletal, cardiac, and visceral muscle tissues.

1. *Skeletal* or *striated voluntary muscle*—attaches to bones; microscopic view shows cells have cross striations (Fig. 2-8); contractions are controlled voluntarily.
2. *Cardiac* or *striated involuntary muscle*—composes wall of heart; cells have cross striations (Fig. 2-9); contractions ordinarily cannot be controlled.
3. *Visceral* or *nonstriated (smooth) involuntary muscle*—helps form walls of blood vessels, intestines, and various other tube-shaped structures; cells appear smooth, that is, without cross striations (Fig. 2-10); contractions ordinarily cannot be controlled. In recent years many individuals have learned some voluntary control of smooth muscle contractions by using biofeedback devices.

All three types of muscle tissue specialize in contraction, a function that produces many kinds of movements in and of the body.

Nervous tissue

Nervous tissue is discussed on p. 84.

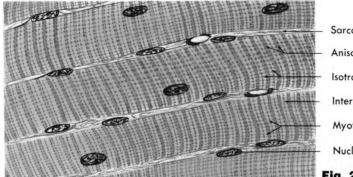

Sarcolemma

Anisotropic substance

Isotropic substance

Intermediate line

Myofibrils

Nucleus

Fig. 2-8 A, Skeletal or striated voluntary muscle tissue. (Courtesy Dr. C. R. McMullen, Department of Biology, South Dakota State University.)

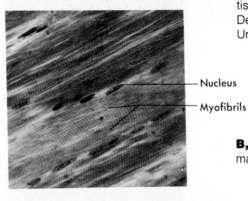

Nucleus

Myofibrils

B, Skeletal (striated) muscle, higher magnification.

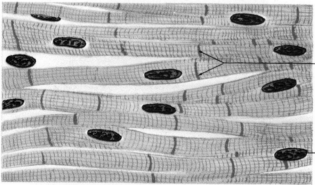

Intercalated discs

Fig. 2-9 A, Cardiac or striated involuntary muscle tissue. (Courtesy Dr. C. R. McMullen, Department of Biology, South Dakota State University.)

Nucleus

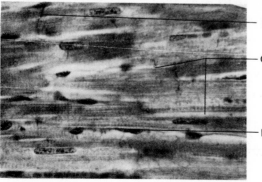

Intercalated disk

Cardiac muscle cells

B, Cardiac muscle, higher magnification.

Nucleus

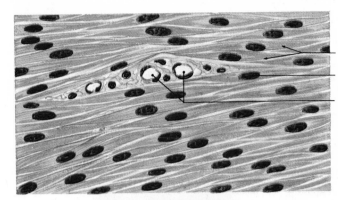

Smooth muscle cells

Nucleus

Blood capillaries

Fig. 2-10 A, Visceral or nonstriated (smooth) involuntary muscle tissue. (Courtesy Dr. C. R. McMullen, Department of Biology, South Dakota State University.)

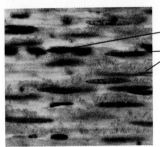

Nucleus

Smooth muscle cells

B, Smooth muscle, higher magnification.

Table 2-2 Tissues

Tissue	Location	Function
Epithelial		
Simple squamous	Alveoli of lungs	Absorption by diffusion of respiratory gases between alveolar air and blood
	Lining of blood and lymphatic vessels	Absorption by diffusion, filtration, and osmosis
Stratified squamous	Surface of lining of mouth and esophagus	Protection
	Surface of skin (epidermis)	Protection
Simple columnar	Surface layer of lining of stomach, intestines, and parts of respiratory tract	Protection; secretion; absorption

Continued.

Table 2-2 Tissues—cont'd		
Tissue	**Location**	**Function**
Connective (most widely distributed of all tissues)		
Areolar	Between other tissues and organs	Connection
Adipose (fat)	Under skin	Protection
	Padding at various points	Insulation; support; reserve food
Dense fibrous	Tendons; ligaments; aponeuroses; scars	Flexible but strong connection
Bone	Skeleton	Support; protection
Cartilage	Part of nasal septum; covering articular surfaces of bones; larynx; rings in trachea and bronchi	Firm but flexible support
	Disks between vertebrae	
	External ear	
Muscle		
Skeletal (striated voluntary)	Muscles that attach to bones	Movement of bones
	Extrinsic eyeball muscles	Eye movements
	Upper third of esophagus	First part of swallowing
Cardiac (striated involuntary)	Wall of heart	Contraction of heart
Visceral (nonstriated involuntary or smooth)	In walls of tubular viscera of digestive, respiratory, and genitourinary tracts	Movement of substances along respective tracts
	In walls of blood vessels and large lymphatic vessels	Change diameter of blood vessels
	In ducts of glands	Movement of substances along ducts
	Intrinsic eye muscles (iris and ciliary body)	Change diameter of pupils and shape of lens
	Arrector muscles of hairs	Erection of hairs (gooseflesh)
Nervous		
	Brain; spinal cord; nerves	Irritability; conduction

Outline summary

Cells

A Composition—cells are composed of protoplasm or "living matter"; protoplasm consists largely of water, proteins, carbohydrates, fats (lipids), and nucleic acids

B Size and shape—human cells vary considerably in size but all are microscopic; cells differ markedly in shape

C Structural parts

1 Cytoplasmic membrane (plasma membrane) forms outer boundary of cell; two layers of phospholipid molecules form framework of membrane; protein molecules lie at outer and inner surfaces of phospholipid layers and many extend through both layers; functions—keeps cells whole and intact; allows certain substances to move through it but keeps others out; chemical messages (for example, hormones) are communicated to cells by binding to certain receptor protein molecules on the membrane's surface

2 Cytoplasm—protoplasm located between cytoplasmic membrane and nucleus; contains following organelles:

a Endoplasmic reticulum—network of connecting sacs and canals extending through cytoplasm from nucleus to cytoplasmic membrane; functions as miniature circulatory system for the cell

b Ribosomes—small spherical structures, some attached to endoplasmic reticulum, some lying free in cytoplasm; synthesize enzymes and other protein compounds, so nicknamed cell's "protein factories"

c Mitochondria—double membranous sacs nicknamed cell's "power plants" because energy-releasing chemical reactions go on inside them

d Lysosomes—membranous-walled sacs nicknamed "digestive bags" and "suicide bags" because they contain enzymes that can digest food compounds or the cell itself

e Golgi apparatus—consists of tiny sacs stacked one upon the other, located near the nucleus; the cell's "carbohydrate-producing and -packaging factory"

3 Nucleus—a small sphere in the central portion of a cell; contains cell's chromosomes

D Functions—major cell functions are the following

1 Movement of substances through cell membranes

a Diffusion—process by which solutes scatter themselves evenly throughout a fluid and by which solutes and water move through a membrane separating two fluids to distribute themselves evenly on both sides of the membrane; diffusion eventually equilibrates concentrations of solutions separated by a membrane

b Osmosis—diffusion of water through a selectively permeable membrane (one that maintains at least one solute concentration gradient); more water osmoses into the fluid with the originally higher solute concentration

c Filtration—movement of water and solutes through a membrane owing to a greater pushing force on one side of the membrane

d Active transport—"uphill" movement of a substance through a living cell membrane, that is, from the solution where a substance's concentration is lower into solution where its concentration is higher; energy for active transport supplied by cellular chemical reactions

2 Cell reproduction

a DNA structure—large molecule shaped like a spiral staircase; sugar, deoxyribose, and phosphate units compose sides of the molecule; base pairs (adenine-thymine or guanine-cytosine) compose "steps"; base pairs always the same but sequence of base pairs differs in different DNA molecules; a gene is a specific sequence of about 1,000 base pairs; genes dictate formation of enzymes and other proteins by ribosomes, thereby indirectly determining cell's structure and functions; in short, genes are heredity-determinants

b DNA replication—process by which each half of a DNA molecule becomes a whole molecule identical to the original DNA molecule; precedes mitosis

c Mitosis—process of cell division that distributes identical chromosomes (DNA molecules) to each of new cells formed when the original cell divides; mitosis enables cells to reproduce their own kind; it makes heredity possible

Tissues

A Epithelial tissue
  1 Simple squamous epithelium—scalelike cells; function is absorption
  2 Stratified squamous epithelium—closely packed cells; function is protection
  3 Simple columnar epithelium—lines stomach, intestine, and parts of respiratory tract
    a Goblet cells secrete mucus
    b Columnar cells specialize in absorption
B Connective tissue
  1 Connects various parts of body
  2 Supports framework of body as a whole and many of its organs
C Muscle tissue
  1 Skeletal muscle—attaches to bones; cells have cross striations; contractions controlled voluntarily
  2 Cardiac muscle—composes wall of heart; cells have cross striations; contractions ordinarily cannot be controlled
  3 Visceral muscle—helps form walls of blood vessels, intestines, and other tubular structures; cells appear smooth; contractions ordinarily cannot be controlled
D Nervous tissue—see p. 84

Review questions

1 Name the three main parts of a cell.
2 Describe the structure and functions of the cytoplasmic membrane. Give another name for this structure.
3 What and where are the following? What functions do they perform?
  endoplasmic reticulum   mitochondria
  Golgi apparatus   ribosomes
  lysosomes
4 Give the full name of the now-famous acid found in cell nuclei.
5 What microscopic structures are composed largely of DNA?
6 Briefly describe DNA's function.
7 What is a gene?

8 Give the scientific names of the structures indicated by the following nicknames:

carbohydrate-producing and -packaging factory   protein factories

digestive bags   suicide bags

power plants

9 How many genes are there in one human cell? Approximately billions? dozens? hundreds? millions? thousands?

10 Name four processes by which substances move through cell membranes.

11 Define each process named in the preceding question.

12 Which of the following processes move solutes through cell membranes: active transport? diffusion? filtration? osmosis?

13 Which of the following processes move water through cell membranes: diffusion? filtration? osmosis?

14 How many inches are in 1 meter?

15 What fractional part of a meter is 1 centimeter? 1 millimeter? 1 micrometer?

16 One inch is approximately equal to how many centimeters? How many millimeters? How many micrometers?

chapter 3

Organs and systems

The words "organ" and "system" have special meanings when applied to the body. An *organ* is a structure made up of two or more kinds of tissues organized in such a way that together they can perform a more complex function than can any one tissue alone. A *system* is a group of organs arranged in such a way that together they can perform a more complex function than can any one organ alone. This chapter discusses a representative organ, the skin, names the systems of the body and the organs that compose them, and briefly describes the functions of each system.

Organs
The skin

The skin is the largest organ in the body and one of the most important. Its area is the surface area of the entire body (about 1.6 to 1.9 square meters in an average-sized adult). Its weight in most adults is 20 pounds or more, making it the body's heaviest organ. Architecturally the skin is a marvel. Consider the incredible number of structures fitted into an area about the size of your little fingernail: several dozen sweat glands, hundreds of nerve endings, yards of tiny blood vessels, numerous oil glands and hairs, and literally thousands of cells.

Structure. Two main layers compose the skin: an outer and thinner layer, the *epidermis*, and an inner and thicker layer, the *dermis* (Fig. 3-1). The epidermis consists of stratified squamous epithelial tissue and the dermis consists of fibrous connective tissue. As you may recall, the cells of stratified epithelial tissue are closely packed together and are arranged in several layers. Only the cells of the innermost layer of the epidermis reproduce themselves. As they increase in number, they move up toward the surface and dislodge the dead horny cells of the outer layer. These flake off by the thousands onto our clothes, into our bathwater, and onto things we handle. Millions of epithelial cells reproduce daily to replace the millions shed—just one example of work our bodies do without our knowing it, even when they seem to be resting.

Have you ever wondered what gives color to your skin or why its color sometimes changes? The answer to the first question is that the epidermis contains deposits of a pigment. Its name is *melanin*. The more melanin, the deeper the skin color is. The amount of melanin in your skin depends first on certain skin color genes you have inherited. Heredity, in a word, determines how dark or light your basic skin color is. Other factors, however, can modify the hereditary effect. The best-known one is sunlight. Prolonged exposure to sunlight darkens the skin because it leads to increased melanin deposits in the epidermis. Skin color frequently changes, however, without a change in the amount of melanin. For example, skin color changes if the volume of blood in the skin changes markedly. If

skin blood volume increases markedly, skin color usually turns pink or even red. If skin blood volume decreases appreciably, skin color changes to a pale, almost white shade. In general, the less abundant the melanin deposits in the skin, the more visible the changes in color with a change in skin blood volume. Conversely, the richer the skin's pigmentation, the less noticeable its color change with a change in skin blood volume.

An interesting structural feature of the dermis or deep layer of the skin is its parallel ridges. Look at the palms of your hands and you will see the ridges and grooves that make possible fingerprinting as a means of identification. Observe in Fig. 3-1 how the epidermis follows the contours of the ridges in the dermis. These develop sometime before birth. Not only is their pattern unique for each individual, but also it never changes except to grow larger—two facts that explain why our fingerprints or footprints positively identify us. Many hospitals identify each newborn baby by footprinting it soon after birth.

Connective tissue composes the dermis. Instead of cells being crowded close together like the epithelial cells of the epidermis, they are scattered far apart, with many fibers in between. Some of the fibers are tough and strong (collagenous or white fibers), and others are stretchable and elastic (elastic fibers). Some of the cells of the dermis store fat. The wrinkles that come with age are believed to be caused partly by

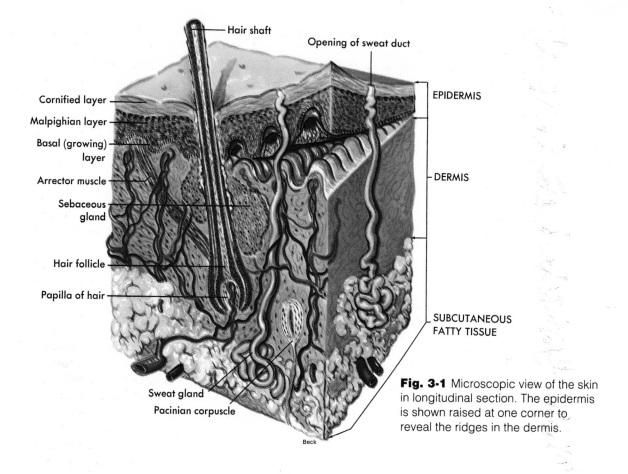

Hair shaft

Opening of sweat duct

EPIDERMIS

Cornified layer

Malpighian layer

Basal (growing) layer

Arrector muscle

Sebaceous gland

DERMIS

Hair follicle

Papilla of hair

SUBCUTANEOUS FATTY TISSUE

Sweat gland

Pacinian corpuscle

Beck

Fig. 3-1 Microscopic view of the skin in longitudinal section. The epidermis is shown raised at one corner to reveal the ridges in the dermis.

a decrease in skin fat and partly by a decrease in the number of elastic fibers. A rich network of capillaries lies in the dermis and also immediately beneath it. The epidermis contains no blood vessels. Surgeons sometimes use the knowledge of this anatomical fact to treat certain skin conditions by cutting such thin shavings from the surface of the skin that they remove only the epidermis. As you would expect, this is a bloodless operation. Examine Fig. 3-1 to find the answers to the following questions: Which layer of the skin—epidermis or dermis—contains hair follicles? oil (sebaceous) glands? sweat glands? Both the blood ves-

sels and the nerves of the skin lie in its dermal layer.

HAIR. If you have ever looked closely at a newborn baby, you probably noticed its soft downy hair. If the baby was born prematurely, you may have found the same type of hair (called "lanugo" from the Latin word that means "down" or "woolliness") all over its tiny body. Coarser hair soon replaces the lanugo of the scalp and eyebrows, but new hair on the rest of the body generally stays delicate and downlike. Hair grows from a cluster of epithelial cells at the bottom of the hair follicle. As long as these cells remain alive, new hair will replace any that is cut or plucked. Contrary to popular belief, frequent cutting or shaving does not make hair grow faster or become coarser. Why? Neither process affects the epithelial cells that form the hairs, since they are embedded in the scalp.

OIL GLANDS. Wherever hairs grow, oil or sebaceous glands also grow. Their tiny ducts open into hair follicles so that their secretions (sebum) lubricate the hair as well as the skin. Someone aptly described sebum as "nature's cold cream." Besides keeping the hair and skin soft and pliant, this natural oil also helps preserve the normal water content of the body by hindering water loss from the skin by evaporation. This in turn helps preserve normal body temperature. Evaporation has a cooling effect, and because the presence of oil on the skin slows down evaporation, it also slows down cooling. Long-distance swimmers apply this knowledge when they coat their bodies with grease before plunging into the water.

A small involuntary muscle attaches to the side of each hair follicle just below its oil gland and extends upward on a slant to the skin. When it contracts, as it often does when we are cold or emotionally upset, it produces several effects. It squeezes on the oil gland with the same result that you would produce if you squeezed on an oil-can—a tiny bit of oil squirts out of each follicle onto the skin. Also, as the muscle contracts, it simultaneously pulls on its two points of attachment—that is, up on a hair follicle but down on a part of the skin. This produces little raised places (goose pimples) between the depressed points of the skin and at the same time pulls the hairs up more or less straight. The name of these muscles, "arrectores pilorum," describes their function; it is Latin for "erectors of the hair." We unconsciously recognize these facts in pictorial expressions such as "I was so frightened my hair stood on end"

and "She was fairly bristling with anger."

SWEAT GLANDS. The pinpoint-sized openings on the skin that you probably call pores are outlets of small ducts from the sweat glands. The glands themselves lie in the subcutaneous tissue just under the dermis, and their ducts spiral up through the epidermis to the surface. There are great numbers of them. For instance, the palms of your hands are estimated to have about 3,000 sweat glands per square inch! The soles of the feet, the forehead, and the axillae (armpits) also contain a great many of them, more than most parts.

Functions. The skin is a vital organ. So crucial to survival are its functions that if about one third of the skin of a healthy young adult is destroyed, death usually results. The skin performs several major functions. It serves as our first line of defense against a multitude of hazards. It protects us against the daily invasion of deadly microbes. It bars the entry of harmful chemicals and prevents the sun's ultraviolet rays from penetrating into the interior of the body. The skin functions as an enormous sense organ. Its millions of nerve endings serve as antennas or receivers for the body, keeping it informed of changes in its environment. The skin helps regulate both the body's temperature and its fluid content by varying the amounts of sweat secreted. Changes in the volume of blood flowing through the skin play an essential role in steadying both body temperature and blood pressure.

Systems

In contrast to cells, which are the smallest structural units of the body, systems are its largest structural units. A *system* consists of a group of organs that work together to perform a more complex function than any one organ can perform alone. Examine Fig.

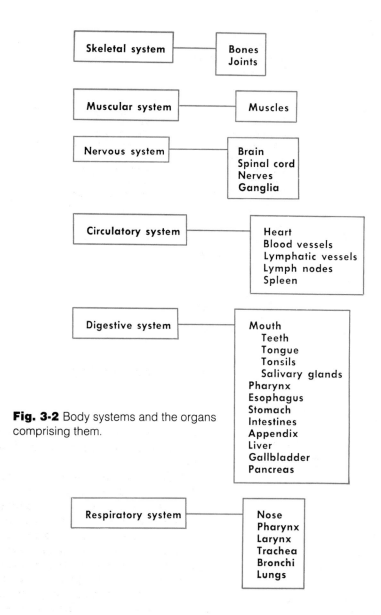

Fig. 3-2 Body systems and the organs comprising them.

3-2 to find the names of the systems. Most anatomists group the organs into nine systems. How many did you find named in Fig. 3-2? Note the names of the organs that make up each of the different systems.

The skeletal system supports the body. The muscular system moves it. Nervous and endocrine systems together provide communication between parts of the body and control of its many functions. The circulatory system provides transportation. The respiratory system absorbs oxygen into the blood from the air in exchange for carbon dioxide, which leaves the blood for the air. The urinary system rids the body of various wastes. And finally, the reproductive systems of male and female together reproduce the human body. Most of the remaining

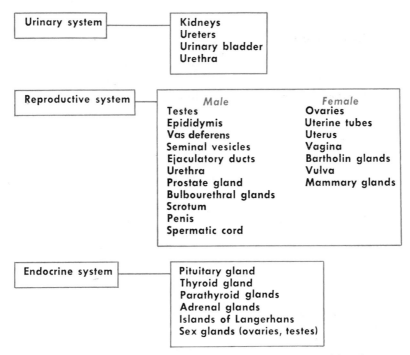

Fig. 3-2, cont'd. Body systems and the organs comprising them.

chapters of this book discuss a single system each, beginning with the skeletal system in Chapter 4.

Outline summary

Organs

A Definition—structures composed of two or more kinds of tissues so organized that together they can perform a more complex function than any one tissue alone

B The skin—a representative organ

  1 Functions

    a Prevents various harmful agents (microbes, chemicals, excess sunlight) from penetrating to interior of body

    b Serves as sense organ; millions of sensory nerve endings in skin

    c Helps regulate body's temperature and its fluid content by varying amounts of sweat secreted and of blood flow through skin

  2 Structure—consists of outer, thinner layer (epidermis) and inner, thicker layer (dermis)

    a Epidermis—composed of stratified squamous epithelium; innermost layer of squamous epithelial cells continually reproduces; new cells move up toward surface, die, and flake off; pigment melanin in epidermis but no blood vessels

    b Dermis—connective tissue layer of skin beneath epidermis; ridges and grooves in dermis form pattern unique to each individual (basis of fingerprinting); blood vessels, nerves, hair roots, oil and sweat glands present in dermis; also, varying numbers of fat cells

Systems

A Names—see Fig. 3-2

B Functions

  1 Skeletal—support

  2 Muscular—movement

  3 Nervous—communication

4 Endocrine—communication
5 Circulatory—transportation
6 Digestive—processes foods
7 Respiratory—exchange of gases between blood and air
8 Urinary—excretes wastes
9 Reproductive—reproduces body

Review questions

1 Give brief definitions of the words "organ" and "system."
2 What organ has the largest surface area and the heaviest weight of any in the body?
3 Describe briefly the vital functions performed by the skin.
4 Name the two main layers that compose the skin.
5 What kind of tissue composes the epidermis? the dermis?
6 Which layer of the skin contains pigment? Name the pigment.
7 Explain the structural feature of the skin that makes possible positive identification of an individual by fingerprinting or footprinting.
8 Which layer of the skin contains hair follicles? oil glands? sweat glands? blood vessels and nerves?
9 What functions do oil and sweat secretion serve?

Systems that form the framework of the body and move it

4 The skeletal system

Functions
Divisions of skeleton
Differences between a man's and a
   woman's skeleton
Differences between a baby's and an
   adult's skeleton
Structure of long bones
Bone formation and growth
Joints (articulations)
Joint changes related to age

5 The muscular system

Muscle tissue
Skeletal muscles (organs)
Specific facts about certain key muscles
Muscle disorders

chapter 4

The skeletal system

Unit two consists of two chapters. This chapter discusses the skeletal system, the system that provides the body with a rigid framework. In this respect the skeletal system functions as steel girders in a building. But it also functions quite differently. Unlike steel girders, bones can move toward or away from each other and in some cases can even move around in circles. Chapter 4 deals with the system that moves bones.

Before studying this chapter, let your imagination go for a few minutes. Think of your bones suddenly turning soft, into a material of the consistency of a piece of liver. Suppose you were standing when this change took place. What do you see happening? Suppose you fell and struck your head. Would you bruise your scalp only or your brain as well? Now change your mental picture. This time think of a single piece of wood cut in the shape of the human body. Try to visualize this structure sitting down, bending over, throwing a ball, or walking across a room. Can you see it moving any of its parts or making any one of the hundreds of movements that you make every day without even giving them a thought?

Functions

The images you have just called up point out the first three of the following functions performed by the skeletal system.

Support

Bones form the body's supporting framework much as steel girders form the supporting framework of our modern buildings.

Protection

Hard, bony "boxes" protect delicate structures enclosed within them. For example, the skull protects the brain. The breastbone and ribs protect vital organs (heart and lungs) and also a vital tissue (red bone marrow, the blood cell–forming tissue).

Movement

Muscles are anchored firmly to bones. As muscles contract and shorten they pull on bones and thereby move them.

Storage

Bones play an important part in maintaining homeostasis of blood calcium, a vital substance. They serve as a safety-deposit box for calcium. When the amount of calcium in blood increases above normal, calcium moves out of the blood into the bones for storage. Conversely, when blood calcium decreases below normal, calcium moves in the opposite direction. It comes out of storage in bones and enters the blood.

Divisions of skeleton

The human skeleton has two divisions: the *axial skeleton* and the *appendicular skeleton*. Bones of the skull, spine, and chest, and the hyoid bone in the neck make up the axial skeleton. The appendicular skeleton consists of the bones of the upper extremities (shoulder girdles, arms, wrists, and hands) and the lower extremities (hip girdles, legs, ankles, and feet). Read Table 4-1. Then locate the various parts of the axial skeleton and the appendicular skeleton in Figs. 4-1 and 4-2.

Table 4-1 Main parts of the skeleton

Axial skeleton	Appendicular skeleton

Skull
  Cranium
	Upper extremities
	  Shoulder girdle

Ear bones	Arms
Face	Wrists
Spine Vertebrae	Hands

Thorax
  Ribs
	Lower extremities
	  Hip girdle

Sternum	Legs
Hyoid bone	Ankles
	Feet

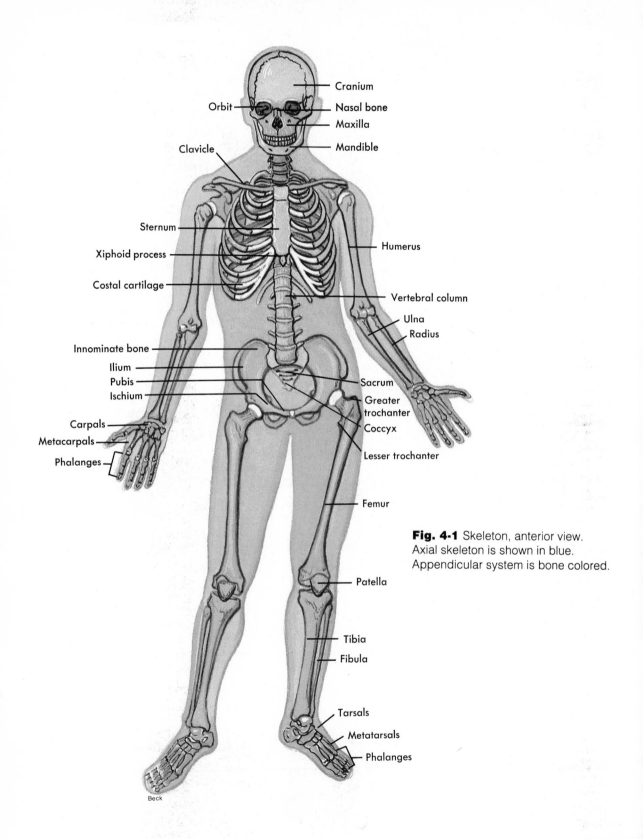

Orbit

Clavicle

Sternum

Xiphoid process

Costal cartilage

Innominate bone

Ilium

Pubis

Ischium

Carpals

Metacarpals

Phalanges

Cranium

Nasal bone

Maxilla

Mandible

Humerus

Vertebral column

Ulna

Radius

Sacrum

Greater trochanter

Coccyx

Lesser trochanter

Femur

Patella

Tibia

Fibula

Tarsals

Metatarsals

Phalanges

Fig. 4-1 Skeleton, anterior view. Axial skeleton is shown in blue. Appendicular system is bone colored.

Beck

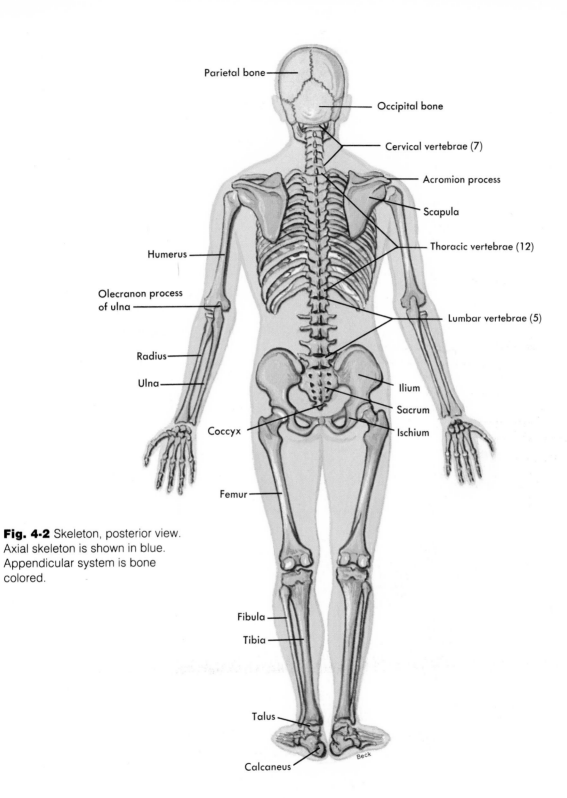

Fig. 4-2 Skeleton, posterior view. Axial skeleton is shown in blue. Appendicular system is bone colored.

Axial skeleton

Skull. The skull consists of eight bones that form the cranium, fourteen bones that form the face, and six tiny bones in the inner ear. You will probably want to learn the names and locations of these bones. These are given in Table 4-2 on p. 54. Find as many of them as you can on Figs. 4-3 and 4-4. Feel their outlines in your own body where possible. Examine them on a skeleton if you have access to one.

"My sinuses give me so much trouble." Have you ever heard this complaint or perhaps uttered it yourself? *Sinuses* are spaces, or cavities, inside some of the cranial bones. Four pairs of them (those in the frontal, maxillary, sphenoid, and ethmoid bones) have openings into the nose and so are referred to as *paranasal sinuses*. Sinuses give trouble when the mucous membrane that lines them becomes inflamed, swollen, and painful. For example, inflamma-

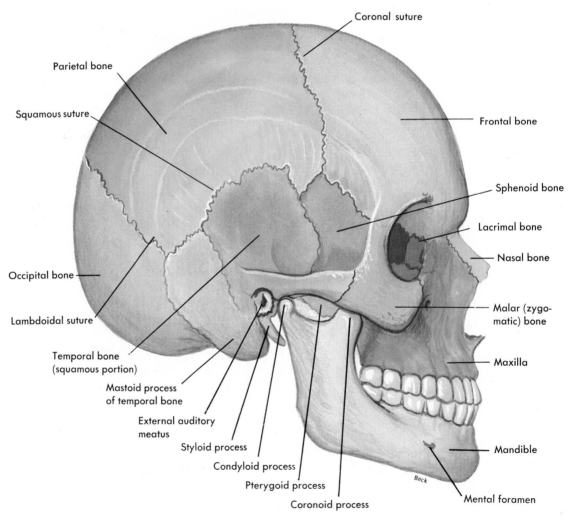

Fig. 4-3 Skull viewed from the right side.

tion in the frontal sinus *(frontal sinusitis)* often starts from a common cold. (The letters "-itis" added to a word mean "inflammation of.")

Spine (vertebral column). The term "vertebral column" may suggest a mental picture of the spine as a single long bone shaped like a column in a building, but this is far from true. The vertebral column consists of a series of separate bones (vertebrae) connected in such a way that they form a flex-ible curved rod. Different sections of the spine have different names: cervical region, thoracic region, lumbar region, sacrum, and coccyx. Read about them in Table 4-2.

Have you ever noticed the four curves in your spine? Your neck and the small of your back curve slightly inward or forward, whereas the chest region of the spine and the lowermost portion curve in the opposite direction. The cervical and lumbar curves of the spine are called concave curves, and

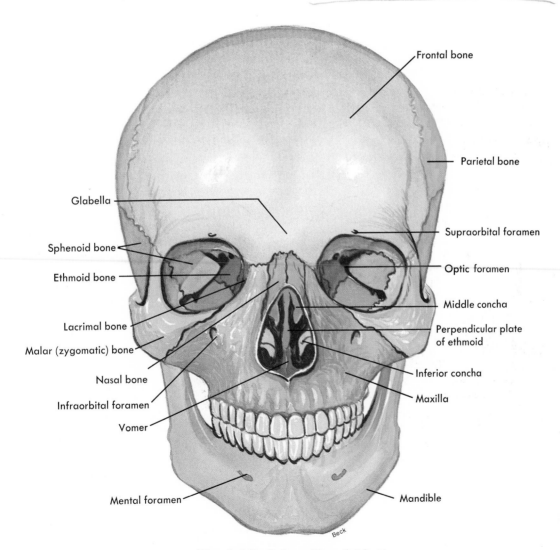

Fig. 4-4 Skull viewed from the front.

the thoracic and sacral curves are called convex curves. This is not true, however, of a newborn baby's spine. It forms a continuous convex curve from top to bottom. Gradually, as the baby learns to hold up his head, a reverse or concave curve develops in his neck (cervical region). And later as the baby learns to stand, the lumbar region of his spine also becomes concave.

The normal curves of the spine serve important functions. They give it enough strength to support the weight of the rest of the body. They also provide the balance necessary for us to stand and walk on two feet instead of having to crawl on all fours. A curved structure has more strength than a straight one of the same size and materials. (The next time you pass a bridge look to see whether or not its supports form a curve.) Clearly the spine needs to be a strong structure. It supports the head balanced on top of it, the ribs and internal organs suspended from it in front, and the hips and legs attached to it below. Poor posture or disease often causes the lumbar curve to become abnormally exaggerated, a condition commonly called "swayback" or technically, *lordosis.* Another abnormal

curvature is *kyphosis,* known to most of us as "hunchback."

Although individual vertebrae are small bones and irregular in shape, they have several well-defined parts. Note, for example, in Fig. 4-5, the body of the lumbar vertebra shown there, its spinous process (or spine), its two transverse processes, and the hole in its center. What is the name of this hole? To feel the tip of the spinous process of one of your own vertebrae, simply bend your head forward and run your fingers down the back of your neck until you feel a projection of bone at shoulder level. This is the tip of the seventh cervical vertebra's long spinous process. The seven cervical vertebrae form the supporting framework of the neck.

Thorax. Twelve pairs of ribs, the sternum (breastbone), and the thoracic vertebrae form the bony cage known as the thorax or chest. Each of the twelve pairs of ribs attaches to a vertebra posteriorly. Also, all the ribs except the lower two pairs attach to the sternum and so have an anterior as well as a posterior anchorage. Look closely at Fig. 4-1 and you can see that the first seven pairs of ribs (sometimes referred to as

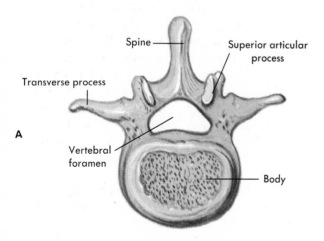

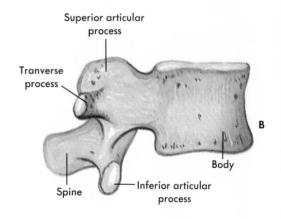

Fig. 4-5 A, Third lumbar vertebra viewed from above. **B,** Third lumbar vertebra viewed from the side.

the "true ribs") attach to the sternum by means of costal cartilage. The eighth, ninth, and tenth pairs of ribs attach to the cartilage of the seventh ribs and are sometimes called "false ribs." The last two pairs of ribs, in contrast, do not attach to any costal cartilage but seem to float free in front—hence their descriptive name, "floating ribs."

Appendicular skeleton

Upper and lower extremities. Well over a hundred bones form the framework of our arms, legs, shoulders, and hips. The upper arm and the upper leg (the thigh) contain only one bone each—the *humerus* in the upper arm and the *femur* in the thigh. Notice in Fig. 4-1 how long these bones are. In fact, the femur is the longest bone in the body and the humerus is the second longest. Take another look at Fig. 4-1. Note the names of the bones in the lower arm and lower leg. Feel the bone running up the thumb side of your lower arm. Is its name radius or ulna? The technical name for the large bone that forms a rather sharp edge along the front of your lower leg is the *tibia*. You may have called this your "shinbone." A slender, somewhat fragile bone, named the *fibula*, lies in the side portion of the lower leg. Fig. 4-6 shows many of the markings on the bones of the lower arm.

The hands have more bones in them for their size than any other part of the body— fourteen finger bones or *phalanges*, five *metacarpal* or palm-of-the-hand bones, and eight *carpal* or top-of-the-hand bones—a total of twenty-seven bones in each hand (Fig. 4-7). This structural fact has great functional importance. It is the presence of many small bones in the hand and wrist and many movable joints between them that makes possible the human hand's high degree of dexterity.

Toe bones have the same name, *pha-langes*, as finger bones. There are the same number of toe bones as finger bones; a fact that might surprise you since toes are so much shorter than fingers. Foot bones comparable to the metacarpals and carpals of the hand have slightly different names. They are called *metatarsals* and *tarsals* in the foot (Fig. 4-8). Just as each hand contains five metacarpal bones, each foot contains five metatarsal bones. But the foot has only seven tarsal bones in contrast to the hand's eight carpals. (A way that helps me remember which of these are foot bones and which are hand bones is to associate tarsal and toes together because they both begin with the letter *t*.)

Feet are for standing on, so certain features of their structure make them able to support the body's weight. The great toe, for example, is considerably more solid and less mobile than the thumb. The foot bones are held together in such a way as to form springy lengthwise and crosswise arches. These provide great supporting strength and a highly stable base. Strong ligaments and leg muscle tendons normally hold the foot bones firmly in their arched positions. Not infrequently, however, the foot ligaments and tendons weaken. The arches then flatten, a condition descriptively called fallen arches or flatfeet (Fig. 4-9).

Two arches extend lengthwise in the foot. One lies on the inside part of the foot and is called the *medial longitudinal arch*. The other lies along the outer edge of the foot and is named the *lateral longitudinal arch*. Another arch extends across the ball of the foot; it is called the *transverse* or *metatarsal arch*.

The shoulder girdle attaches the arms to the trunk, and the *hip girdle* attaches the legs to the trunk. The two clavicles (collarbones) and the two scapulas (shoulderblades) form the shoulder girdle. The two large *innominate bones* (pelvic or hipbones)

Text continued on p. 48.

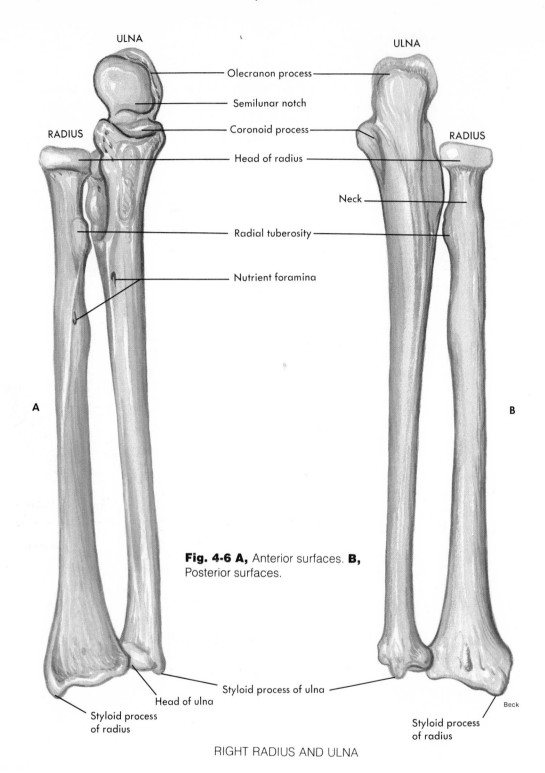

ULNA

Olecranon process

Semilunar notch

RADIUS

Coronoid process

Head of radius

ULNA

RADIUS

Neck

Radial tuberosity

Nutrient foramina

A

B

Fig. 4-6 A, Anterior surfaces. **B,** Posterior surfaces.

Styloid process of ulna

Head of ulna

Styloid process
of radius

Styloid process
of radius

Beck

RIGHT RADIUS AND ULNA

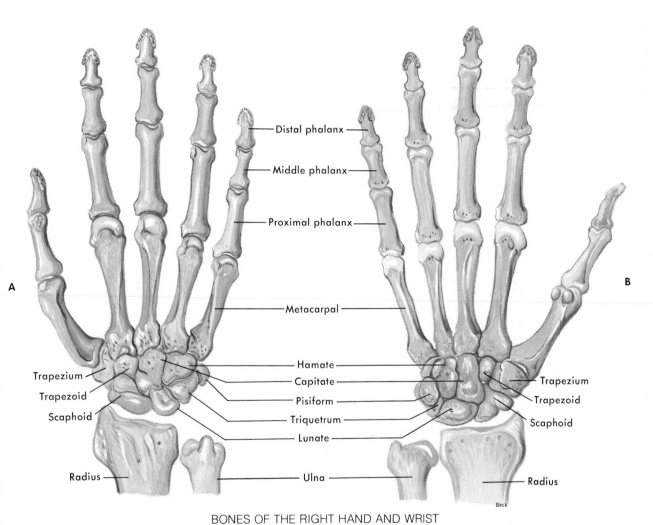

Distal phalanx

Middle phalanx

Proximal phalanx

A

B

Metacarpal

Hamate
Capitate
Pisiform
Triquetrum
Lunate

Trapezium
Trapezoid
Scaphoid

Trapezium
Trapezoid
Scaphoid

Radius

Ulna

Radius

Beck

BONES OF THE RIGHT HAND AND WRIST

Fig. 4-7 A, Dorsal surface. **B,** Palmar surface.

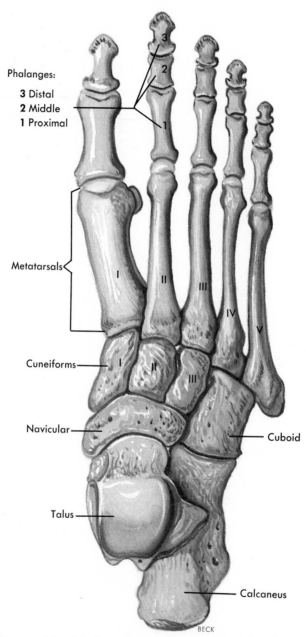

Phalanges:

3 Distal
2 Middle
1 Proximal

Metatarsals

Cuneiforms

Navicular

Cuboid

Talus

Calcaneus

BECK

Fig. 4-8 Bones of the right foot viewed from above. Tarsal bones consist of cuneiform bones, navicular bone, talus, cuboid bone, and calcaneus. Compare the names and numbers of foot bones shown here with those of the hand bones shown in Fig. 4-7.

Fig. 4-9 Flatfoot results when there is a weakening of tendons and ligaments attached to the tarsal bones. Downward pressure by the weight of the body gradually flattens out the normal arch of bones.

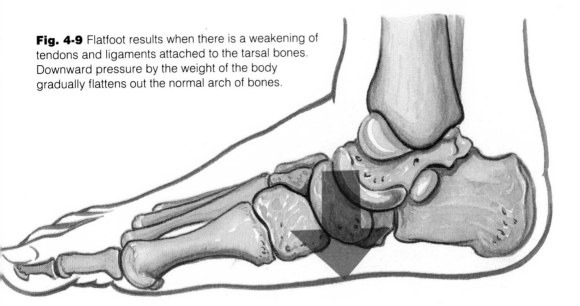

Fig. 4-10 High heels throw the weight forward causing the heads of the metatarsals to bear most of the body's weight.

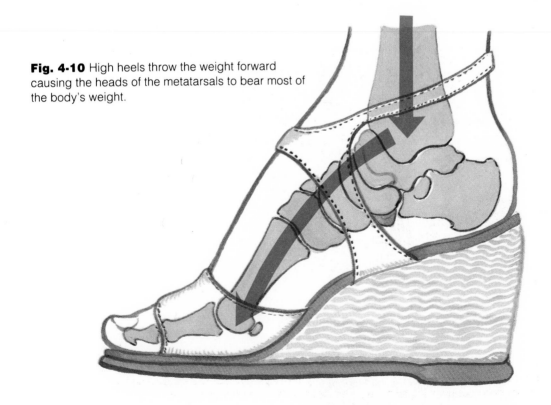

compose the hip girdle. In a baby's body each innominate bone consists of three bones—the ilium, ischium, and pubis. Later these bones grow together to become one bone in an adult.

Differences between a man's and a woman's skeleton

A man's skeleton and a woman's skeleton differ in several ways. Were you to examine a male skeleton and a female skeleton placed side by side, you would probably notice first the difference in their sizes. Most male skeletons are larger than most female skeletons—a structural fact that seems to have no great functional importance. Structural differences between the male and female hipbones, however, do have functional importance. The female pel-

vis is made so that the body of a baby can be cradled in it before birth and can pass through it during birth journey. Although the individual male hipbones (innominate bones) are generally larger than the individual female hipbones, together the male hipbones form a narrower structure than do the female hipbones. A man's pelvis is shaped something like a funnel, but a woman's pelvis has a broader, shallower shape, more like a basin. (Incidentally, the word pelvis means "basin.") Another difference is that the *pelvic inlet*, or brim, is normally much wider in the female than in the male. Figs. 4-11 and 4-12 show this difference clearly. In these figures you can also see how much wider the angle is at the front of the female pelvis where the two pubic bones join than it is in the male.

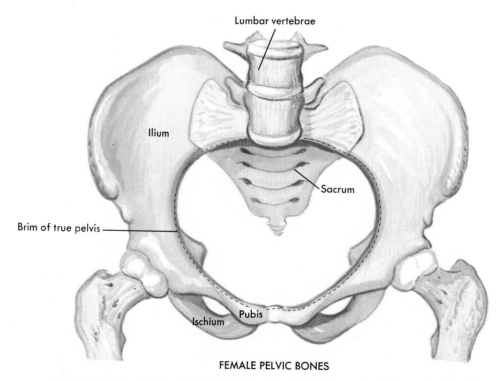

FEMALE PELVIC BONES

Fig. 4-11 Compare the shape of the female pelvis shown here with that of the male pelvis shown in Fig. 4-12.

Differences between a baby's and an adult's skeleton

Have you ever noticed that a newborn baby's head seems larger for the size of its body than an adult's head does? This is one of several differences between a baby's skeleton and an adult's skeleton. A newborn's head is about one fourth as long as its whole body, whereas an adult's head is only about one eighth of its total height. A baby's face appears small and its cranium large; an adult's face and cranium look nearly the same size. A baby's chest is round; an adult's is oval.

At birth the skeleton is unfinished, since some of the bones still consist partly of cartilage or fibrous tissue. Familiar examples of this are the soft spots (fontanels) in a baby's head. Another example is the cartilage between the shaft of a long bone and its ends. This is known as the *epiphyseal cartilage,* and it remains as long as bones are still growing. Epiphyseal cartilage can be seen in x-ray films—an important fact for a doctor who wants to know if a child is going to grow any more. No epiphyseal cartilage means no more growth in bone length.

Structure of long bones

Figs. 4-13 and 4-14 will help you learn the names of the main parts of a long bone. Identify each of the following:
   1. *Diaphysis* or shaft—a hollow tube made of hard compact bone, hence, a rigid and strong structure that is light

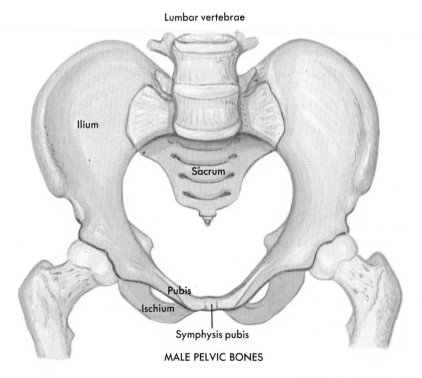

Lumbar vertebrae

Ilium

Sacrum

Pubis

Ischium

Symphysis pubis

MALE PELVIC BONES

Fig. 4-12 Note the narrower width of this male pelvis compared with the female pelvis shown in Fig. 4-11.

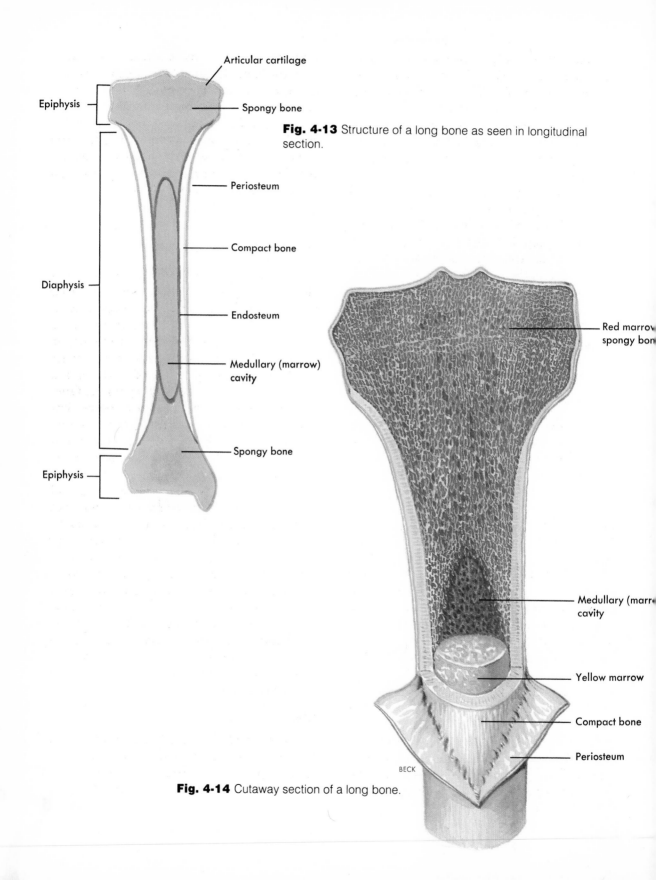

Articular cartilage

Epiphysis

Spongy bone

Fig. 4-13 Structure of a long bone as seen in longitudinal section.

Periosteum

Compact bone

Diaphysis

Endosteum

Medullary (marrow) cavity

Epiphysis

Spongy bone

Red marrow spongy bone

Medullary (marrow) cavity

Yellow marrow

Compact bone

BECK

Periosteum

Fig. 4-14 Cutaway section of a long bone.

enough in weight for easy movement

2. *Medullary cavity*—the hollow area inside the diaphysis of a bone; contains soft yellow bone marrow
3. *Epiphyses* or the ends of the bone—red bone marrow fills in small spaces in the spongy bone composing the epiphyses
4. *Articular cartilage*—a thin layer of cartilage covering each epiphysis; serves much the same purpose that a small rubber cushion would if it were placed over the ends of bones where they come together to form a joint
5. *Periosteum*—a strong fibrous membrane that covers a long bone except at its joint surfaces where it is covered by articular cartilage

Bone formation and growth

When the skeleton first forms in a baby before its birth, it consists not of bones but of cartilage and fibrous structures shaped like bones. Gradually these become transformed into real bones by a process too complex for discussion here. It involves special bone-forming and bone-destroying cells and the laying down of calcium salts in the ground substance of the newly forming bones. This calcification process is what makes bones as "hard as bone." As bones grow, the combined action of the bone-destroying and bone-forming cells sculptures the bones into their adult shapes.

A long bone grows from a center in its diaphysis and from other centers in the layers of cartilage that separate each of its epiphyses from the diaphysis. The diaphysis grows in both directions toward both epiphyses, and both epiphyses grow toward the diaphysis. As long as any epiphyseal cartilage remains, this two-way growth continues. It ceases when all the epiphyseal carti-

lage has become transformed into bone. Doctors sometimes use this knowledge to determine whether a child is going to grow any more. They have the child's wrist x-rayed, and if it shows a layer of epiphyseal cartilage, they know that he will grow some more. If, however, it shows no epiphyseal cartilage, they know that he is as tall as he is ever going to be.

Joints (articulations)

Every bone in the body, except one, connects to at least one other bone. In other words, every bone but one forms a joint with some other bone. (The exception is the hyoid bone in the neck, to which the tongue anchors.) Most of us probably never think much about our joints unless something goes wrong with them and they do not function properly. Then their tremendous importance becomes painfully clear. Joints perform two functions: they hold our bones together securely, and at the same time they make it possible for movement to occur between the bones—between most of them, that is. Without joints we could not move our arms, legs, or any other parts. Our bodies would, in short, be rigid, immobile hulks. Try, for example, to move your arm at your shoulder joint in as many directions as you can. Try to do the same thing at your elbow joint. Now examine the shape of the bones at each of these joints on a skeleton or in Figs. 4-1 and 4-2. Do you see why you cannot move your arm at your elbow in nearly as many directions as you can at your shoulder?

Kinds of joints

One method classifies joints according to the degree of movement they allow into three types: synarthroses (no movement), amphiarthroses (slight movement), and diarthroses (free movement). Differences in the structure of joints account for differences in the degree of movement they make possible.

Synarthroses. A synarthrosis is a joint in which fibrous connective tissue grows between the articulating (joining) bones holding them close together. The joints between cranial bones are synarthroses, commonly called sutures.

Amphiarthroses. An amphiarthrosis is a joint in which cartilage connects the articulating bones. The symphysis pubis, the joint between the two pubic bones, is an amphiarthrosis, as are the joints between the bodies of the vertebrae.

Diarthroses. Fortunately, most of our joints by far are diarthroses. All such joints allow considerable movement—sometimes in many directions and sometimes in only one or two directions.

STRUCTURE OF DIARTHROSES (FREELY MOVABLE JOINTS). Freely movable joints are all made alike in certain ways. All of them

have a joint capsule, a joint cavity, and a layer of cartilage over the ends of two joining bones (Fig. 4-15). The *joint capsule* is made of the body's strongest and toughest material—fibrous connective tissue—and is lined with smooth, slippery synovial membrane. The capsule fits over the ends of the two bones something like a sleeve. Because it attaches firmly to the shaft of each bone to form its covering (called the periosteum; *peri* means ''around'' and *osteon* means ''bone''), the joint capsule holds the bones together securely but at the same time permits movement at the joint. The structure of the joint capsule, in other words, helps make possible the joint's function.

Ligaments (cords or bands made of the same strong fibrous connective tissue as the joint capsule) also grow out of the periosteum and lash the two bones together even more firmly.

The layer of *articular cartilage* over the joint ends of bones acts like a rubber heel on a shoe—it absorbs jolts. The *synovial*

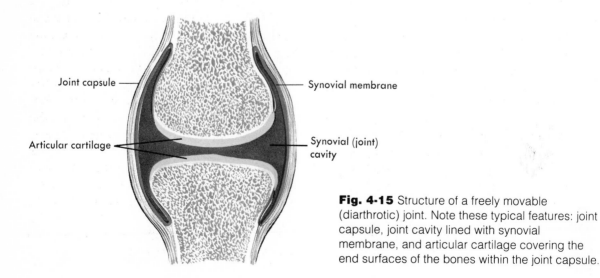

Joint capsule — Synovial membrane

Articular cartilage — Synovial (joint) cavity

Fig. 4-15 Structure of a freely movable (diarthrotic) joint. Note these typical features: joint capsule, joint cavity lined with synovial membrane, and articular cartilage covering the end surfaces of the bones within the joint capsule.

membrane secretes a lubricating fluid (synovial fluid), which allows easier movement with less friction.

There are several types of diarthroses, namely, ball-and-socket, hinge, pivot, saddle, and certain others. Because they differ somewhat in structure, they differ also in their possible range of movement. In a *ball-and-socket joint*, a ball-shaped head of one bone fits into a concave socket of another bone. Shoulder and hip joints, for example, are ball-and-socket joints. Of all the joints in our bodies, these permit the widest range of movements. Think for a moment how many ways you can move your upper arms. You can flex them (move them forward), you can extend them (move them backward), you can abduct them (move them away from the sides of your body), and you can adduct them (move them back down to your sides). You can also circumduct them (move them around so as to describe a circle with your hands).

Hinge joints, like the hinges on a door, allow movements in only two directions, namely, flexion and extension. Flexion is bending a part, extension is straightening it out. Elbow and knee joints and the joints in the fingers are hinge joints.

Pivot joints are those in which a small projection of one bone pivots in an arch of another bone. For example, a projection of the axis, the second vertebra in the neck, pivots in an arch of the atlas, the first vertebra in the neck. This rotates the head, which rests on the atlas.

Only one pair of *saddle joints* exists in the body—between the metacarpal bone of each thumb and a carpal bone of the wrist (the name of this carpal bone is the trapezium). Because the articulating surfaces of these bones are saddle-shaped, they make possible the human thumb's great mobility, a mobility no animal's thumb possesses. We can flex, extend, abduct, ad-

duct, and circumduct our thumbs, and most important of all, we can move our thumbs to touch the tip of any one of our fingers. (This movement is called opposing the thumb to the fingers.) Without the saddle joints at the base of each of our thumbs, we could not do such a simple act as picking up a pin or grasping a pencil between thumb and forefinger.

Joints between the bodies of the vertebrae are diarthroses. These joints make it possible to flex the trunk forward or sideways and even to circumduct and rotate it. Strong ligaments connect the bodies of the vertebrae and fibrous disks lie between them. The central core of these intervertebral disks consists of a pulpy, elastic substance that loses some of its resiliency with age. It may then be compressed by sudden exertion or injury, with fragments protruding into the spinal canal and pressing on spinal nerve routes of the spinal cord. Severe pain results. Medical terminology calls this a herniated disk; popular language calls it a "slipped disk." (See Fig. 4-16.)

Joint changes related to age

Advancing years take their toll on the joints by causing degenerative changes (*osteoarthritis*) in them. This is especially true for the weight-bearing joints and for the joints of overweight individuals. The articular cartilages undergo degenerative changes in osteoarthritis, and excess bone grows along the joint edges of bones. These accumulations of bone are called "spurs" and "marginal lipping." The presence of osteoarthritis undoubtedly accounts for many of the complaints made by older people, such as, "My joints are getting so stiff I even have trouble getting up out of a chair" or "I'm not nearly as spry as I used to be; I can't move around so easily."

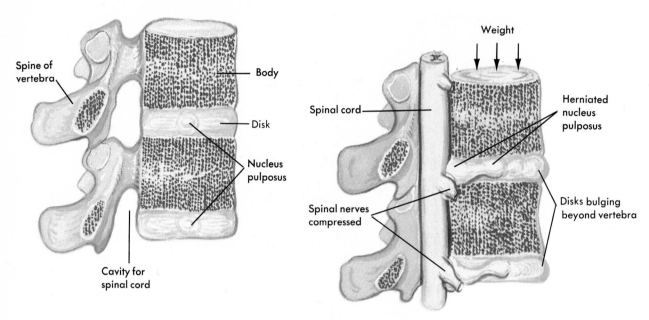

Fig. 4-16 Sagittal section of vertebrae.

Table 4-2 Bones of the skeleton

Name	Number	Description
Cranial bones		
Frontal	1	Forehead bone; also forms front part of floor of cranium and most of upper part of eye sockets; cavity inside bone above upper margins of eye sockets (orbits) called *frontal sinus;* lined with mucous membrane
Parietal	2	Form bulging topsides of cranium
Temporal	2	Form lower sides of cranium; contain *middle* and *inner ear structures; mastoid sinuses* are mucosa-lined spaces in *mastoid process,* the protuberance behind ear; *external auditory canal* is tube leading into temporal bone
Occipital	1	Forms back of skull; spinal cord enters cranium through large hole *(foramen magnum)* in occipital bone
Sphenoid	1	Forms central part of floor of cranium; pituitary gland located in small depression sphenoid called sella turcica *(Turkish saddle)*
Ethmoid	1	Complicated bone that helps form floor of cranium, side walls and roof of nose and part of its middle partition (nasal septum), and part of orbit; contains honeycomb-like spaces, the *ethmoid sinuses; superior* and *middle turbinate bones* (conchae) are projections of ethmoid bone; form "ledges" along side wall of each nasal cavity

Table 4-2 Bones of the skeleton—cont'd

Name	Number	Description
Face bones		
Nasal	2	Small bones that form upper part of bridge of nose
Maxillary	2	Upper jawbones; also help form roof of mouth, floor, and side walls of nose and floor of orbit; large cavity in maxillary bone is *maxillary sinus*
Zygoma (malar)	2	Cheek bones; also help form orbit
Mandible	1	Lower jawbone
Lacrimal	2	Small bone; helps form medial wall of eye socket and side wall of nasal cavity
Palatine	2	Form back part of roof of mouth and floor and side walls of nose and part of floor of orbit
Inferior turbinate	2	Form curved "ledge" along inside of side wall of nose, below middle turbinate
Vomer	1	Forms lower, back part of nasal septum
Ear bones		
Malleus	2	Malleus, incus, and stapes are tiny bones in middle ear cavity in temporal bone; malleus means "hammer"—shape of bone
Incus	2	Incus means "anvil"—shape of bone
Stapes	2	Stapes means "stirrup"—shape of bone
Hyoid bone	1	U-shaped bone in neck at base of tongue
Vertebral column		
Cervical vertebrae	7	Upper seven vertebrae, in neck region; first cervical vertebra called *atlas;* second called *axis*
Thoracic vertebrae	12	Next twelve vertebrae; ribs attach to these
Lumbar vertebrae	5	Next five vertebrae; those in small of back
Sacrum	1	In child, five separate vertebrae; in adult, fused into one
Coccyx	1	In child, three to five separate vertebrae; in adult, fused into one
Thorax		
True ribs	14	Upper seven pairs; attach to sternum by way of *costal cartilages*
False ribs	10	Lower five pairs; lowest two pairs do not attach to sternum, therefore, called *floating ribs;* next three pairs attach to sternum by way of costal cartilage of seventh ribs
Sternum	1	Breastbone; shaped like a dagger; piece of cartilage at lower end of bone called *xiphoid process*

Continued.

Table 4-2 Bones of the skeleton—cont'd

Name	Number	Description
Upper extremities		
Clavicle	2	Collarbones; only joints between shoulder girdle and axial skeleton are those between each clavicle and sternum
Scapula	2	Shoulder bones; scapula plus clavicle forms *shoulder girdle; acromion process*—tip of shoulder that forms joint with clavicle; *glenoid cavity*—arm socket
Humerus	2	Upper arm bone
Radius	2	Bone on thumb side of lower arm
Ulna	2	Bone on little finger side of lower arm; *olecranon process*—projection of ulna known as elbow or "funny bone"
Carpal bones	16	Irregular bones at upper end of hand; anatomical wrist
Metacarpals	10	Form framework of palm of hand
Phalanges	28	Finger bones; three in each finger, two in each thumb
Lower extremities		
Pelvic bones	2	Hipbones; *ilium*—upper flaring part of pelvic bone; *ischium*—lower back part; *pubic bone*—lower front part; acetabulum—hip socket; *symphysis pubis*— joint in midline between two pubic bones; pelvic inlet—opening into *true pelvis,* or pelvic cavity; if pelvic inlet is misshapen or too small, infant skull cannot enter true pelvis for natural birth
Femur	2	Thigh or upper leg bones; *head of femur*—ball-shaped upper end of bone; fits into acetabulum
Patella	2	Kneecap
Tibia	2	Shinbone; *medial malleolus*—rounded projection at lower end of tibia commonly called inner anklebone
Fibula	2	Long slender bone of lateral side of lower leg; *lateral malleolus*—rounded projection at lower end of fibula commonly called outer anklebone
Tarsal bones	14	Form heel and back part of foot; anatomical ankle
Metatarsals	10	Form part of foot to which toes attach; tarsal and metatarsal bones so arranged that they form three arches in foot: *inner longitudinal arch* and *outer longitudinal arch,* both of which extend from front to back of foot, and transverse or *metatarsal arch* that extends across foot
Phalanges	28	Toe bones; three in each toe, two in each great toe
Total	206	

Table 4-3 Identification of bone markings

Bone	Marking	Description
Frontal	Supraorbital margin	Arched ridge just below eyebrows
	Frontal sinuses	Cavities inside bone just above supraorbital margin; lined with mucosa; contain air
Temporal	Mastoid process	Protuberance just behind ear
	Mastoid air cells	Air-filled mucosa-lined spaces within mastoid process
	External auditory meatus (or canal)	Opening into ear and tube extending into temporal bone
	Zygomatic process	Projection that articulates with malar (or zygomatic) bone
	Mandibular fossa	Oval depression anterior to external auditory meatus; forms socket for condyle of mandible
Occipital	Foramen magnum	Hole through which spinal cord enters cranial cavity
	Condyles	Convex, oval processes on either side of foramen magnum; articulate with depressions on first cervical vertebra
Sphenoid	Body	Hollow, cubelike central portion
	Sella turcica (or Turkish saddle)	Saddle-shaped depression on upper surface of sphenoid body; contains pituitary gland
	Sphenoid sinuses	Irregular air-filled mucosa-lined spaces within central part of sphenoid
Ethmoid	Horizontal (cribriform) plate	Olfactory nerves pass through numerous holes in this plate
	Crista galli	Meninges attach to this process
	Perpendicular plate	Forms upper part of nasal septum
	Ethmoid sinuses	Honeycombed, mucosa-lined air spaces within lateral masses of bone
	Superior and middle turbinates (conchae)	Help to form lateral walls of nose
Mandible	Body	Main part of body; forms chin
	Condyle (or head)	Part of each ramus that articulates with mandibular fossa of temporal bone
	Alveolar process	Teeth set into this arch
Maxilla	Alveolar process	Arch containing teeth
	Maxillary sinus or antrum of Highmore	Large air-filled mucosa-lined cavity within body of each maxilla; largest of sinuses
Special features of skull	Sutures	Immovable joints between skull bones
	1 Sagittal	1 Joint between two parietal bones
	2 Coronal	2 Joint between parietal bones and frontal bone
	3 Lambdoidal	3 Joint between parietal bones and occipital bone
	Fontanels	"Soft spots" where ossification incomplete at birth; allow some compression of skull during birth; also important in determining position of head before delivery; six such areas located at angles of parietal bones

Continued.

Table 4-3 Identification of bone markings—cont'd

Bone	Marking	Description
	1 Anterior (or frontal)	**1** At intersection of sagittal and coronal sutures (juncture of parietal bones and frontal bone); diamond shaped; largest of fontanels; usually closed by 1½ years of age
	2 Posterior (or occipital)	**2** At intersection of sagittal and lambdoidal sutures (juncture of parietal bones and occipital bone); triangular in shape; usually closed by second month
	Wormian bones	Small islands of bones within suture
Sternum	Body	Main central part of bone
	Manubrium	Flaring, upper part
	Xiphoid process	Projection of cartilage at lower border of bone
Scapula (Fig. 4-2)	Spine	Sharp ridge running diagonally across posterior surface of shoulderblade
	Acromion process	Slightly flaring projection at lateral end of scapular spine; may be felt as tip of shoulder; articulates with clavicle
	Coracoid process	Projection on anterior from upper border of bone; may be felt in groove between deltoid and pectoralis major muscles about 1 inch below clavicle
	Glenoid cavity	Arm socket
Humerus	Head	Smooth, hemispherical enlargement at proximal end of humerus
Ulna (Fig. 4-6)	Olecranon process	Elbow
	Styloid process	Sharp protuberance at distal end; can be seen from outside on posterior surface
Radius (Fig. 4-6)	Head	Disk-shaped process forming proximal end of radius; articulates with capitulum of humerus and with radial notch of ulna
	Styloid process	Protuberance at distal end on lateral surface (with forearm supinated as in anatomical position)
Innominate (hip)	Ilium	Upper, flaring portion
	Ischium	Lower, posterior portion
	Pubic bone or pubis	Medial, anterior section
	Acetabulum	Hip socket; formed by union of ilium, ischium, and pubis
	Iliac crests	Upper, curving boundary of ilium
	Anterosuperior spine	Prominent projection at anterior end of iliac crest; can be felt externally as "point" of hip
	Ischial tuberosity	Large, rough, quadrilateral process forming posterior part of ischium; in erect sitting position body rests on these tuberosities

Table 4-3 Identification of bone markings—cont'd

Bone	Marking	Description
	Symphysis pubis	Cartilaginous, amphiarthrotic joint between pubic bones
	Obturator foramen	Large hole in anterior surface of os coxae; formed by pubis and ischium; largest foramen in body
	Pelvic brim (or inlet)	Boundary of opening leading into true pelvis; size and shape of this inlet has great obstetrical importance since if any of its diameters too small, infant skull cannot enter true pelvis for natural birth
	True (or lesser) pelvis	Space below pelvic brim; true "basin" with bone and muscle walls and muscle floor; pelvic organs located in this space
	False (or greater) pelvis	Broad, shallow space above pelvic brim inlet; name "false pelvis" is misleading since this space is actually part of abdominal cavity, not pelvic cavity
Femur (Fig. 4-1)	Head	Rounded, upper end of bone; fits into acetabulum
	Greater trochanter	Protuberance located interiorly and laterally to head
	Lesser trochanter	Small protuberance located inferiorly and medially to greater trochanter
Tibia (Fig. 4-1)	Condyles	Large, rounded bulges at distal end of femur; one on medial and one on lateral surface
	Crest	Sharp ridge on anterior surface
	Medial malleolus	Rounded downward projection at distal end of tibia; forms prominence on inner surface of ankle
Fibula (Fig. 4-2)	Lateral malleolus	Rounded prominence at distal end of fibula; forms prominence on outer surface of ankle
Tarsals (Fig. 4-7)	Calcaneus	Heel bone
	Talus	Uppermost of tarsals; articulates with tibia and fibula; boxed in by medial and lateral malleoli
	Longitudinal arches	Tarsals and metatarsals so arranged as to form arch from front to back of foot
	1 Inner	**1** Formed by calcaneus, navicular, cuneiforms, and three medial metatarsals
	2 Outer	**2** Formed by calcaneus, cuboid, and two lateral metatarsals
	Transverse (or metatarsal) arch	Metatarsals and distal row of tarsals (cuneiforms and cuboid) so articulated as to form arch across foot; bones kept in two arched positions by means of powerful ligaments in sole of foot and by muscles and tendons

Outline summary

Functions

A Supports and gives shape to body
B Protects internal organs
C Helps make movements possible
D Stores calcium

Divisions of skeleton

Skeleton composed of the following two main divisions and their subdivisions:
A Axial skeleton
  1 Skull
  2 Spine
  3 Thorax
  4 Hyoid bone
B Appendicular skeleton
  1 Upper extremities, including shoulder girdle
  2 Lower extremities, including hip girdle
C Location and description of bones—see Figs. 4-1 to 4-8, 4-11, and 4-12 and Table 4-2

Differences between a man's and a woman's skeleton

A In size—male skeleton generally larger
B In shape of pelvis—male pelvis deep and narrow, female pelvis broad and shallow
C In size of pelvic inlet—female pelvic inlet generally wider, normally large enough for baby's head to pass through
D In pubic angle—angle between pubic bones of female generally wider

Differences between a baby's and an adult's skeleton

A Baby's head proportionately longer—about one-fourth total length of body; length of adult head about one-eighth total height of body
B Baby's face smaller than cranium; adult's face and cranium about the same size
C Baby's chest is round, adult's is oval
D Parts of baby's skeleton consist of cartilage or fibrous tissue such as "soft spots" in skull

Structure of long bones

See Figs. 4-8 and 4-9

Bone formation and growth

Joints (articulations)

A Kinds of joints
  1 Synarthroses (no movement)—fibrous connective tissue grows between articulating bones; for example, sutures of skull
  2 Amphiarthroses (slight movement)—cartilage connects articulating bones; for example, symphysis pubis
  3 Diarthroses (free movement)—most joints of the body belong to this class
    a Structures of freely movable joints—joint capsule and ligaments hold joining bones together but permit movement at joint
    b Articular cartilage—covers joint ends of bones and absorbs joints
    c Synovial membrane—lines joint capsule and secretes lubricating fluid
    d Joint cavity—space between joint ends of bones
B Types of freely movable joints—ball-and-socket, hinge, pivot, saddle, and certain others

Joint changes related to age

Osteoarthritis, condition in which articular cartilages undergo degenerative changes and excess bone (called "spurs" and "marginal lipping") grows along joint edges, hampering movement

New words

arthritis	lordosis
articulation	lumbar
appendicular skeleton	osteoarthritis
axial skeleton	periosteum
cervical	rickets
diaphysis	sinus
epiphyses	sinusitis
kyphosis	synovial membrane
ligament	thorax

Review questions

1 What functions does the skeletal system perform for the body?

2 Give the anatomical name of the following:

bone on little finger	lower jaw bone
side of lower arm	shoulder-blade
collarbone	thigh bone
finger bones	toe bones
forehead bone	upper arm bone

3 Name the small bones in the middle ear.

4 What functions do joints perform?

5 Describe the structure of a freely movable joint.

6 Locate and briefly describe each of the following bones:

clavicle	phalanges
ethmoid	radius
femur	scapula
frontal	sphenoid
humerus	sternum
ilium	tarsals
innominate	tibia
mandible	turbinates
metacarpals	zygoma
parietal	ulna
patella	

7 Describe one functionally important difference between a male and a female adult skeleton.

8 Describe some differences between an infant and an adult skeleton.

9 Is it possible to tell whether a child is going to grow any taller? If so, how can a doctor tell this?

10 What structural changes frequently occur in older people's joints?

chapter 5

The muscular system

The muscular system consists of more than 500 muscles that move us about in many ways, varying in complexity from blinking an eye or smiling to climbing a mountain or ski jumping. Not many of our body structures can claim as great an importance for happy useful living as can our voluntary muscles, and only a few can boast of greater importance for life itself. A great deal is known about muscles—enough, in fact, to fill several books. The plan for this chapter is to investigate first the different types of muscle tissue, then to note some general facts about the structure and function of skeletal muscles, next to present some specific facts about certain key muscles, and finally to consider some muscle disorders.

Muscle tissue

If you weigh 120 pounds, about 50 pounds of your weight comes from your muscles, the "red meat" attached to your bones. Under the microscope these muscles appear as bundles of fine threads with many crosswise stripes. Each fine thread is a muscle cell or, as it is usually called, a *muscle fiber*. This type of muscle tissue has three names: *striated muscle*—because of its cross stripes or striae, *skeletal muscle*—because it attaches to bone, and *voluntary muscle*—because its contractions can be controlled voluntarily.

Besides skeletal muscle, the body also

contains two other kinds of muscle tissue, cardiac muscle and nonstriated, smooth or involuntary muscle. Cardiac muscle, as its name suggests, composes the bulk of the heart. As you can see in Fig. 5-1, cardiac muscle cells branch frequently. Nonstriated or smooth muscle lacks the cross stripes or striae seen in skeletal muscle. It has a "smooth," even appearance when viewed with a microscope. It is called involuntary because we normally do not have control over its contractions. Smooth or involuntary muscle forms an important part of blood vessel walls and of many hollow internal organs (viscera) such as the gut. (See Fig. 5-1.)

Muscle cells specialize in the function of contraction, or shortening. Every movement we make is produced by contractions of skeletal muscle cells. Contractions of cardiac muscle cells keep the blood circulating through its vessels, and smooth muscle contractions do many things, for instance, move food into and through the stomach and intestines and make a major contribution to the maintenance of normal blood pressure.

Skeletal muscles (organs)
Structure

A skeletal muscle is an organ composed mainly of striated muscle cells and connective tissue. Most skeletal muscles attach to two bones that have a movable joint between them. In other words, most muscles extend from one bone across a joint to another bone. Also, one of the two bones moves less easily than the other. The muscle's attachment to this more stationary bone is called its *origin*. Its attachment to the more movable bone is called the muscle's *insertion*. The rest of the muscle (all of it except its two ends) is called the *body* of the muscle. *Tendons* anchor muscles firmly to bones. Made of dense fibrous connective tissue in the shape of heavy cords, tendons have great strength. They do not tear or pull away from bone easily. Yet any emergency-room nurse or doctor sees many tendon injuries—both severed tendons and tendons torn loose from bones.

Small fluid-filled sacs called *bursas* lie between some tendons and the bones beneath them. These small sacs are made of connective tissue and lined with *synovial membrane*. The synovial membrane secretes a slippery, lubricating fluid (synovial fluid) that fills the bursa. Like a small, flexible cushion, a bursa makes it easier for a tendon to slide over a bone when the tendon's muscle shortens. *Tendon sheaths* enclose certain tendons. Because these tube-shaped structures are also lined with synovial membrane and moistened with synovial fluid, they, like the bursas, facilitate movement. Both bursas and tendon sheaths can become inflamed. Inflammation of a bursa—a relatively common ailment, particularly in older people—is called *bursitis*.

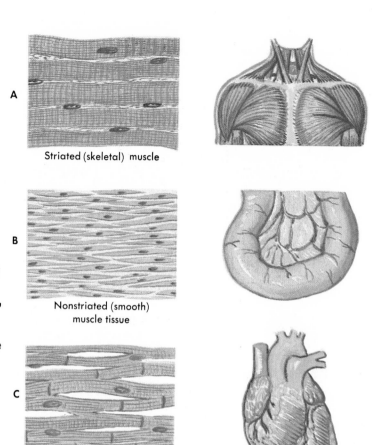

Fig. 5-1 Muscle tissues of the human body. **A,** Striated (voluntary), or skeletal, muscle tissue; left, a microscopic view showing cross striations and multinuclei in each cell; right, a macroscopic view of skeletal muscle organs. **B,** Nonstriated (involuntary), smooth, muscle tissue; left, a microscopic view; right, a loop of intestine, one of many internal organs whose walls contain smooth muscle. **C,** Branching, or cardiac, muscle tissue; left, a microscopic view showing cross striations and branching cells; right, the heart, the only organ made of cardiac muscle tissue.

Striated (skeletal) muscle

Nonstriated (smooth) muscle tissue

Branching (cardiac) muscle tissue

Inflammation of a tendon sheath is *tenosynovitis.*

Functions

Muscles move bones by pulling on them. Because the length of a skeletal muscle becomes shorter as its fibers contract, the bones to which the muscle attaches move closer together. As a rule, only the insertion bone moves. The shortening of the muscle pulls the insertion bone toward the origin bone. The origin bone stays put, holding firm, while the insertion bone moves toward it (Fig. 5-2). One tremendously important function of skeletal muscle contractions, therefore, is to produce movements.

Remember this rule: a muscle's insertion bone moves toward its origin bone. It can help you understand muscle actions.

Another important fact to remember about the functioning of skeletal muscles is this: they generally work in teams, not singly. Several muscles contract at the same time to produce almost any movement you can think of. Of the muscles contracting simultaneously, the one mainly responsible for producing the movement is called the *prime mover* for that movement. The other muscles that help produce the movement are called *synergists.* For example, the prime mover for extension of the lower leg is the rectus femoris muscle of the thigh.

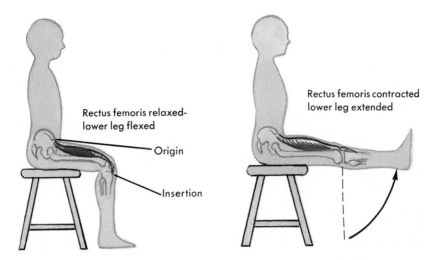

Rectus femoris relaxed-
lower leg flexed

Origin

Insertion

Rectus femoris contracted
lower leg extended

Fig. 5-2 Diagram illustrates the principle that when a muscle contracts, it pulls its insertion bone toward its origin bone. For example, when the rectus femoris muscle (part of the quadriceps femoris muscle group) contracts, it pulls its insertion bone, the tibia, toward its origin, the hipbone. This straightens the knee joint and extends the lower leg as shown.

The synergists that help the rectus femoris bring about this movement are various other muscles located on the front of the thigh.

An important fact to remember about muscles is that they cannot produce movements all by themselves. Other structures must function with them—bones and joints, for example. Skeletal muscles (except for a few in the face) bring about movements by pulling on bones across movable joints. But before a skeletal muscle can contract and pull on a bone to move it, the muscle must first be stimulated by nerve impulses. Muscles also have a constant need for oxygen and food. If food materials or oxygen supplies are cut off, the muscle will soon *fatigue* or weaken because of a lack of energy. Energy for all kinds of work done by all kinds of body cells, you will recall, comes from the cells' catabolism of foods. Catabolism uses oxygen, and in order for cells to receive their vital oxygen supply, the respiratory system must move oxygen from the air into the blood. The circulatory system must then deliver the oxygen to the body cells. In short, the respiratory, circulatory, nervous, muscular, and skeletal systems all play essential parts in producing normal movements. This fact has great practical importance. For example, a person might have perfectly normal muscles and still not be able to move normally. He might have a nervous system disorder that shuts off impulses to certain skeletal muscles and thereby paralyzes them. Multiple sclerosis is one great enemy that acts in this way, but so do some other conditions—a hemorrhage in the brain, a brain tumor, or a spinal cord injury, to mention a few. Skeletal system disorders, arthritis especially, can have disabling effects on movement. Muscle functioning, then, depends on the functioning of many other parts of the body. This fact illustrates a principle repeated often in this book. It can be simply stated: no part of the body lives by or for itself alone. Each part depends on all other parts for its healthy survival. Each part contributes something to the healthy survival of all other parts.

Contraction of a skeletal muscle does not always produce movement. Sometimes it increases the tension within a muscle but does not decrease the length of the muscle. When the muscle does not shorten, no movement results. One type of contraction that produces no movement is known as *isometric contraction.* The word "isometric" comes from Greek words that mean "equal measure." In other words, a muscle's length during an isometric contraction and during relaxation is about equal. Although muscles do not shorten (and therefore do not produce movement) during isometric contractions, the tension within them increases. Because of this, repeated isometric contractions tend to make muscles grow larger and stronger. Hence the popularizing in recent years of isometric exercises as great muscle builders. The type of muscle contraction that produces movement is *isotonic contraction.* Try this simple experiment to see for yourself the difference between isometric and isotonic contractions. Watch the front of your upper arm as you bend your arm at the elbow and move your hand up toward your shoulder. As you make this movement, can you see the muscle in your upper arm become shorter and more bulging? Now place one hand on the undersurface of a heavy table or desk top, and, still watching your upper arm, push up on the table top with all your strength. Did your upper arm muscle (biceps brachii is its name) shorten this time as it contracted? Did it produce movement? What kind of contraction did you perform this time—isometric or isotonic? Answer question 7 on p. 77.

Muscle tone, or tonic contraction, is another type of skeletal muscle contraction that produces no movement. Because relatively few of a muscle's fibers shorten at one time in a tonic contraction, the muscle as a whole does not shorten. Consequently, tonic contractions do not move any body parts. What they do is hold them in position. Or, using synonyms for the words "tonic contractions and position," muscle tone maintains *posture.* Posture and movements are the two great functions of skeletal muscles. Both contribute greatly to our healthy survival. Good posture means that body parts are held in the positions that favor best function. It means positions that balance the distribution of weight and that therefore put the least strain on muscles, tendons, ligaments, and bones. To have good posture in a standing position, for example, you must stand with your head and chest held high, your chin, abdomen, and buttocks pulled in, and your knees bent slightly.

To judge for yourself how important good posture is, consider some of the effects of poor posture. Besides detracting from appearance, poor posture makes a person tire more quickly. It puts an abnormal pull on ligaments, joints, and bones and therefore often leads to deformities. Poor posture crowds the heart, making it harder for it to contract. Poor posture crowds the lungs, decreasing their breathing capacity.

Skeletal muscle tone maintains posture by counteracting the pull of gravity. Gravity tends to pull the head and trunk down and forward, but the tone in certain back and neck muscles pulls just hard enough in the opposite direction to overcome the force of gravity and hold the head and trunk erect. The tone in thigh and leg muscles puts just enough pull on thigh and leg bones to counteract the pull of gravity on them that would otherwise collapse the hip and knee joints and cause us to fall in a heap.

Perhaps the most important function of

skeletal muscle is movement itself. This basic function is even more important than good posture for health. Most of us believe that "exercise is good for us" even if we have no idea what or how many specific benefits can come from it. The benefits of exercise are almost legion. Professional journals carry articles about exercising and so do many popular magazines and books.* Some of the good consequences of regular, properly practiced exercise are greatly improved muscle tone, better posture, more efficient heart and lung functioning, less fatigue, and, last but not least, looking and feeling better.

In addition to movement and posture, skeletal muscles have a third important function—heat production. We must maintain our body temperature within a very narrow range. If body temperature varies even a degree or two from normal—98.6° F—we become sick. Muscle cells, like all cells, produce heat by the process of catabolism. But because skeletal muscle cells are both highly active and numerous, they produce a major share of total body heat. Skeletal muscle contractions therefore constitute one of the most important parts of the mechanism that helps us maintain a constant body temperature.

Sick people just naturally move about less than well people. Medically speaking, they become less mobile or even immobile. Too little mobility inevitably has bad effects on virtually all body structures and functions. These effects often threaten survival and sometimes lead to death. Good nursing care, however, can do much to hold off or lessen the damages of diminished activity. Later chapters describe many harmful effects of immobility.

TYPES OF MOVEMENTS PRODUCED BY SKELETAL MUSCLE CONTRACTIONS. Muscles can move some body parts in several directions and others in only two directions. As mentioned on p. 51, this depends largely on the shapes of the bones and the type of freely movable joint. Some of the movements we make most often are flexion, extension, abduction, and adduction.

Suppose that you bend one of your arms at the elbow or bend a leg at the knee, or suppose that you bend over forward at the waist or bend your head forward in prayer. With each of these movements you will have flexed some part of the body—the lower arm, the lower leg, the trunk, and the head, respectively. Most flexions are movements commonly described as bending. One exception to this rule is flexion of the upper arm. Flexing the upper arm means moving it forward from the chest, as shown in Fig. 5-3. *Flexion* is defined as a movement that makes the angle between two bones at their joint smaller than it was at the beginning of the movement (Fig. 5-6).

Extensions are opposite, or antagonistic, actions to flexions. Thus extensions are movements that make joint angles larger rather than smaller. Extensions are straightening or stretching movements rather than bending movements. Extending the lower arm, for example, is straightening it out at the elbow joint, as shown in Fig. 5-7, A. Extending the upper arm is stretching it out and back from the chest as shown in Fig. 5-4. What movement would you make to extend your lower leg? to extend your thigh or upper leg? Look at Figs. 5-7, B, and 5-8, A, to check your answers. If you straighten your back to stand tall or stretch backward from your waist, are you flexing or extending your trunk?

*Cooper, K. K.: The new aerobics, New York, 1978, M. Evans & Co., p. 191; Fixx, J. F.: The complete book of running, New York, 1977, Random House, p. 314; Pollock, M. L., Wilmore, J. H., and Fox, S. M.: Health and fitness through physical activity, New York, 1978, John Wiley & Sons, p. 357.

Pectoralis major muscle

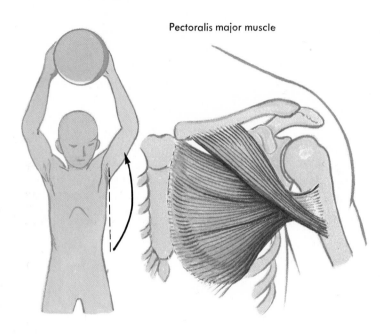

Fig. 5-3 When the pectoralis major muscle (shown on the figure at the right) contracts, it flexes the upper arm at the shoulder joint (figure at the left). In what bone must the pectoralis major insert if it moves the upper arm?

Latissimus dorsi muscle

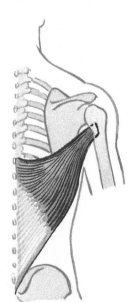

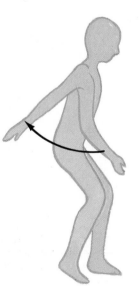

Fig. 5-4 When the latissimus dorsi muscle (shown on the figure at the right) contracts, it extends the upper arm at the shoulder joint (figure at the left). In what bone must the latissimus dorsi insert if it moves the upper arm?

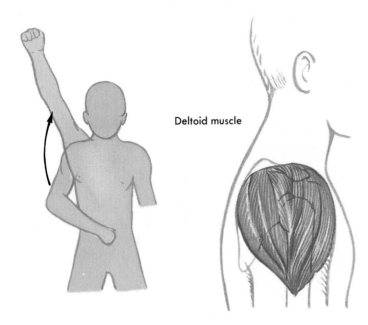

Deltoid muscle

Fig. 5-5 When the deltoid muscle (shown on the figure at the right) contracts, it abducts the upper arm at the shoulder joint (figure at the left). In what bone must the deltoid insert to produce this movement?

Hamstring muscles

Biceps brachii muscle

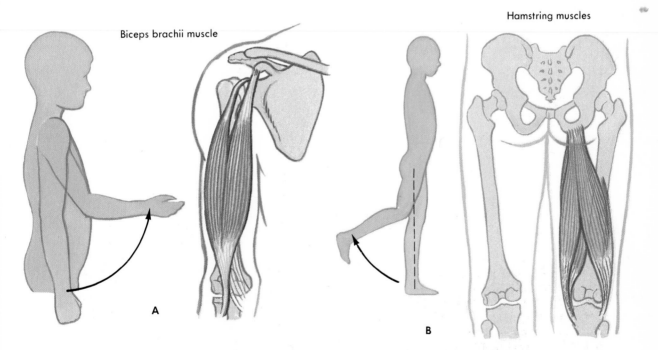

A

B

Fig. 5-6 Flexion of the lower arm and lower leg. **A,** When the biceps brachii muscle (shown at the right) contracts, it flexes the lower arm at the elbow joint (shown at the left). **B,** When the hamstring muscles (shown at the right) contract, they flex the lower leg at the knee joint (shown at the left).

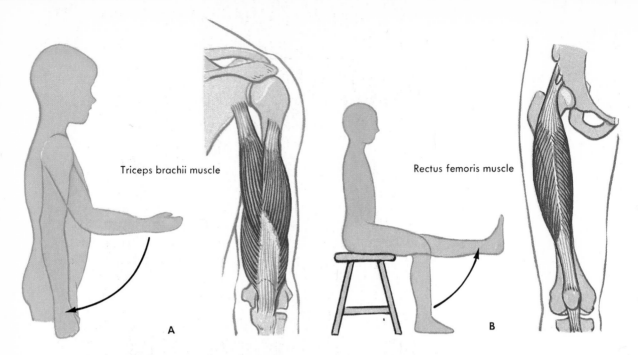

Fig. 5-7 Extension of the lower arm and lower leg. **A,** When the triceps brachii muscle (shown at the right) contracts, it extends the lower arm at the elbow joint (shown at the left). **B,** When the rectus femoris muscle (part of the quadriceps femoris muscle group) (shown at the right) contracts, it extends the lower leg at the knee joint (shown at the left).

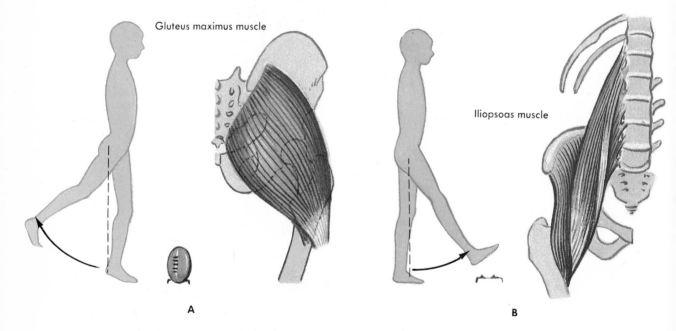

Fig. 5-8 Flexion and extension of the thigh. **A,** When the gluteus maximus muscle (shown at the right) contracts, it extends the thigh at the hip joint (shown at the left). **B,** When the iliopsoas muscle (shown at the right) contracts and the femur serves as its insertion, it flexes the thigh at the hip joint (shown at the left).

Abduction means moving a part away from the midline of the body, such as moving the arms out to the sides (Fig. 5-5).

Adduction means moving a part toward the midline, such as bringing the arms down to the sides.

Figs. 5-9 to 5-12 show the outer layer of muscles known as the *superficial muscles.* *Deep muscles* lie under most of these.

Specific facts about certain key muscles

Flexors (that is, muscles that flex different parts of the body) produce many of the movements used in walking, sitting, swimming, typing, and a host of other activities. Extensors also function in these activities but perhaps play their most important role in maintaining upright posture. Study Table 5-1 and Figs. 5-9 to 5-12 to learn the names of some of the prime flexors, extensors, abductors, and adductors of the body. Consult Table 5-2 to learn their origins and insertions. Keep in mind that muscles move bones, and the bones they move are their insertion bones. You might use this information this way—if you learn that a certain muscle flexes the upper arm, you can deduce that it has its insertion on what bone? If a muscle is listed as an extensor of the lower leg, on what bone or bones must it insert? Answer review questions 9 and 10 on p. 77.

Table 5-1 Muscles grouped according to function

Part moved	Flexors	Extensors	Abductors	Adductors
Upper arm	Pectoralis major	Latissimus dorsi	Deltoid	Pectoralis major and latissimus dorsi contracting together
Lower arm	Biceps brachii	Triceps brachii	None	None
Thigh	Iliopsoas	Gluteus maximus	Gluteus medius and minimus	Adductor group
Lower leg	Hamstrings	Quadriceps femoris group	None	None
Foot	Tibialis anterior	Gastrocnemius and soleus	Peroneus longus	Tibialis anterior
Trunk	Iliopsoas and rectus femoris	Erector spinae (sacrospinalis)	Psoas major and quadratus lumborum	Psoas major and quadratus lumborum

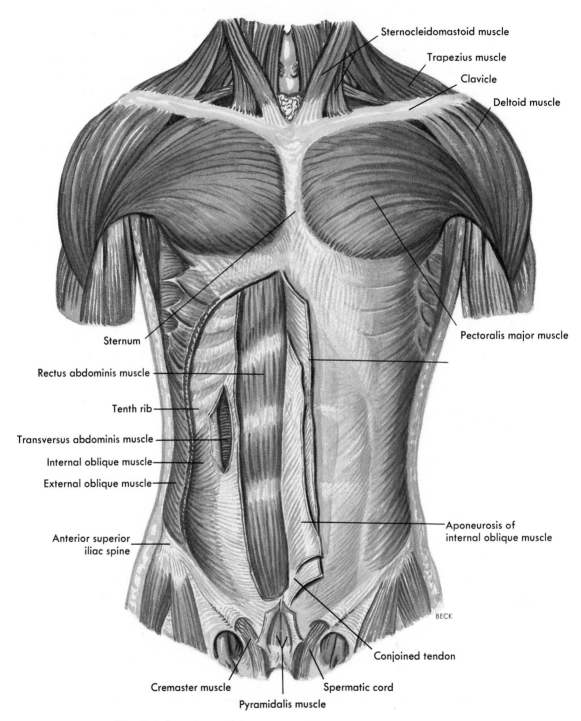

Labels on the figure:

Sternocleidomastoid muscle
Trapezius muscle
Clavicle
Deltoid muscle
Pectoralis major muscle
Sternum
Rectus abdominis muscle
Tenth rib
Transversus abdominis muscle
Internal oblique muscle
External oblique muscle
Anterior superior iliac spine
Aponeurosis of internal oblique muscle
Conjoined tendon
Cremaster muscle
Spermatic cord
Pyramidalis muscle

BECK

Fig. 5-9 Outer layer of the muscles of the anterior surface of the trunk.

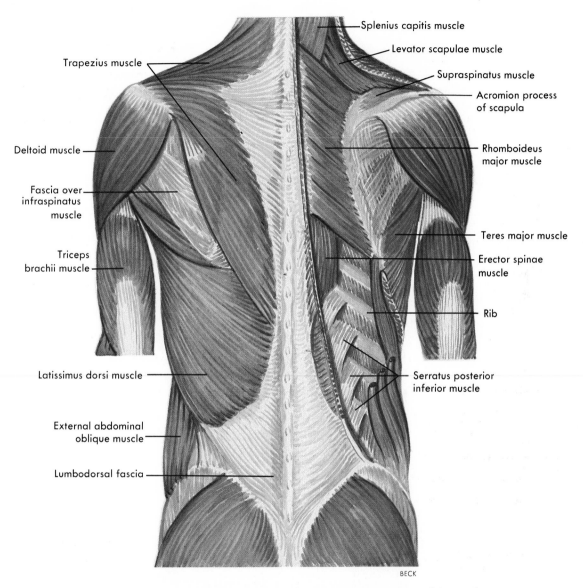

Splenius capitis muscle

Levator scapulae muscle

Supraspinatus muscle

Acromion process of scapula

Trapezius muscle

Rhomboideus major muscle

Deltoid muscle

Fascia over infraspinatus muscle

Teres major muscle

Triceps brachii muscle

Erector spinae muscle

Rib

Latissimus dorsi muscle

Serratus posterior inferior muscle

External abdominal oblique muscle

Lumbodorsal fascia

BECK

Fig. 5-10 Outer layer of the muscles of the posterior surface of the trunk.

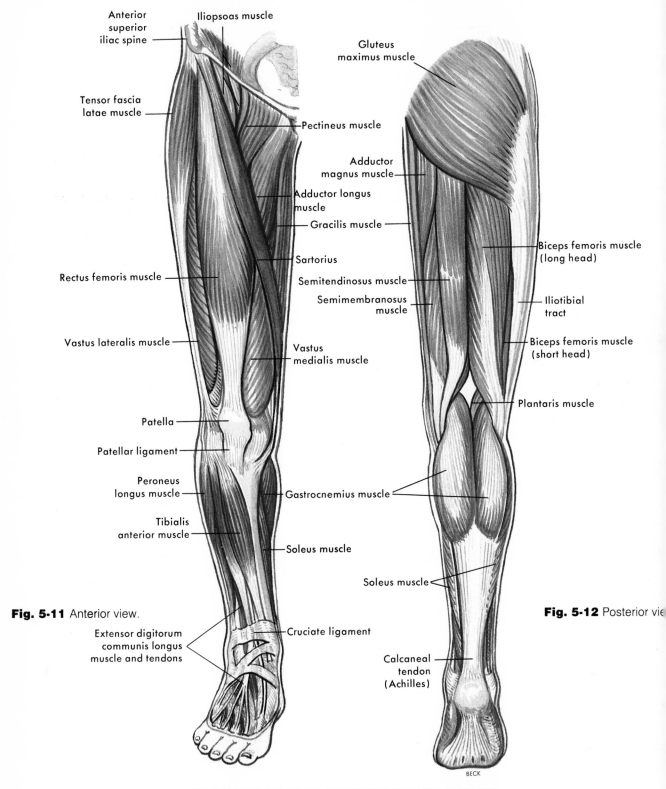

Anterior superior iliac spine

Iliopsoas muscle

Gluteus maximus muscle

Tensor fascia latae muscle

Pectineus muscle

Adductor magnus muscle

Adductor longus muscle

Gracilis muscle

Biceps femoris muscle (long head)

Rectus femoris muscle

Sartorius

Semitendinosus muscle

Semimembranosus muscle

Iliotibial tract

Vastus lateralis muscle

Vastus medialis muscle

Biceps femoris muscle (short head)

Plantaris muscle

Patella

Patellar ligament

Peroneus longus muscle

Gastrocnemius muscle

Tibialis anterior muscle

Soleus muscle

Soleus muscle

Fig. 5-11 Anterior view.

Fig. 5-12 Posterior vie

Extensor digitorum communis longus muscle and tendons

Cruciate ligament

Calcaneal tendon (Achilles)

BECK

SUPERFICIAL MUSCLES OF THE RIGHT THIGH AND LEG

Table 5-2 Muscle functions, origins, and insertions

Muscle	Function	Insertion	Origin
Pectoralis major	Flexes *upper arm* Helps adduct *upper arm*	Humerus	Sternum Clavical Upper rib cartilages
Latissimus dorsi	Extends *upper arm* Helps adduct *upper arm*	Humerus	Vertebrae Ilium
Deltoid	Abducts *upper arm*	Humerus	Clavicle Scapula
Biceps brachii	Flexes *lower arm*	Radius	Scapula
Triceps brachii	Extends *lower arm*	Ulna	Scapula Humerus
Iliopsoas	Flexes *trunk*	Ilium Vertebrae	Femur
Iliopsoas	Flexes *thigh*	Femur	Ilium Vertebrae
Gluteus maximus	Extends *thigh*	Femur	Ilium Sacrum Coccyx
Gluteus medius	Abducts *thigh*	Femur	Ilium
Gluteus minimus	Abducts *thigh*	Femur	Ilium
Adductors	Adducts *thigh*	Femur	Pubic bone
Hamstring group	Flexes *lower leg* Helps extend *thigh*	Tibia Fibula	Ischium Femur
Quadriceps femoris group, including rectus femoris	Extends *lower leg* Helps flex *thigh*	Tibia	Ilium Femur

Muscle disorders

Perhaps the best known disease of muscles is *muscular dystrophy*. It is a long-lasting disease whose main characteristics are progressive wasting and weakening of the muscles.

A person with *paralysis* cannot contract his muscles when he wants to; his skeletal muscles do not respond to his will. This is not because there is anything wrong with his muscles but because there is disease or injury of his brain or spinal cord or nerves so that they cannot activate his muscles.

Muscle atrophy is muscle shrinkage, a decrease in muscle size caused by disuse. Many conditions can cause atrophy, such as having a leg in a cast, being a bed patient for a long time, or paralysis.

Muscle hypertrophy is the opposite of atrophy; that is, it is an increase in size resulting from increased use. Skeletal muscles may hypertrophy as a result of exercise. The heart frequently hypertrophies from overwork.

Another disease characterized by muscular weakness is *myasthenia gravis*. The cause of this sometimes swiftly fatal disease is still not established. It appears, however, that the disease may involve a defect in the body's immune system, which leads to a lack of communication between nerves and muscles.* The result is progressive muscular weakness and paralysis. In severe cases death may occur because of paralysis of the respiratory muscles.

Outline summary

Muscle tissue

A Structure and types
  1 Striated muscles; also called skeletal muscle or voluntary muscle
  2 Branching muscle; also called cardiac muscle

*Target of the error in myasthenia, Science News **115**:102, 1979.

  3 Nonstriated muscle; also called smooth muscle, visceral muscle, or involuntary muscle
B Function—muscle cells specialize in contraction (shortening)

Skeletal muscles (organs)

A Structure
  1 Composed mainly of striated muscle cells (fibers) and connective tissue
  2 Most muscles extend from one bone across movable joint to another bone
  3 Parts of a skeletal muscle
    a Origin—attachment to relatively stationary bone
    b Insertion—attachment to bone that moves
    c Body—main part of muscle
  4 Muscles attach to bones by tendons—strong cords of fibrous connective tissue; some tendons enclosed in synovial-lined tubes, lubricated by synovial fluid; tubes called tendon sheaths; inflammation of tendon sheaths—tenosynovitis
  5 Bursas—small synovial-lined sacs containing small amount synovial fluid; located between some tendons and underlying bones
B Functions
  1 Muscles produce movement—as muscle contracts, it pulls insertion bone nearer origin bone; movement occurs at joint between origin and insertion
    a Groups of muscles usually contract to produce a single movement
      (1) Prime mover—muscle whose contraction is mainly responsible for producing a given movement
      (2) Synergists—muscles whose contractions help the prime mover produce a given movement
    b Types of movements produced by skeletal muscle contractions
      (1) Flexion—making angle at joint smaller
      (2) Extension—making angle at joint larger
      (3) Abduction—moving a part away from midline
      (4) Adduction—moving a part toward midline
    c Normal muscle functioning depends on

normal functioning of various other structures, notably nerves and joints

  d Not all muscle contractions produce movements; isometric contractions increase the tension in muscles without producing movements; isotonic contractions produce movements; tonic contractions, or muscle tone, produce no movement but increase firmness (tension) of muscles that maintain posture, that is, the position of parts of body

  e Mobility (exercise) absolutely essential for healthy survival—necessary for maintaining muscle tone, for normal functioning of heart and lungs, for normal structure and functioning of bones and joints, and so on

2 Posture

  a Posture means position of body parts

  b Good posture important for many reasons—for example, to prevent fatigue and bone and joint deformities

  c Skeletal muscles maintain posture by counteracting the pull of gravity—by maintaining tone (partial contraction)

3 Heat production

  a Body temperature must be maintained within a narrow range of normal (98.6° F)

  b Muscle cells produce body heat by catabolism of foodstuffs as they shorten; because muscle cells are highly active and numerous, they produce largest percent of body heat and constitute one of the most important parts of the mechanism that helps us maintain constant body temperature

Specific facts about certain key muscles

See Tables 5-1 and 5-2

Muscle disorders

A Muscular dystrophy—progressive wasting and weakening of muscles

B Paralysis—loss of ability to produce voluntary movements

C Atrophy—decrease in muscle size

D Hypertrophy—increase in muscle size

E Myasthenia gravis—progressive muscular weakness and paralysis

New words

abduction	isotonic
adduction	muscular dystrophy
atrophy	muscle tone
body of muscle	myasthenia gravis
bursa	origin (of muscle)
bursitis	paralysis
contraction	posture
extension	striated
fatigue	synovial fluid
flexion	synovial membrane
hypertrophy	tendon
insertion (of muscle)	tendon sheath
isometric	tenosynovitis

Review questions

1 Compare the three kinds of muscle tissue as to location, microscopic appearance, and nerve control.

2 Explain why skeletal muscle functions are so important. What are the general functions of skeletal muscle?

3 Explain the following terms and give an example of each:

  flexion   abduction
  extension   adduction

4 Explain how skeletal muscles, bones, and joints work together to produce movements.

5 Why can a spinal cord injury be followed by muscle paralysis?

6 Can a muscle contract very long if its blood supply is shut off? Give a reason for your answer.

7 What two kinds of muscle contractions do not produce movement?

8 The correct term to substitute in the expression "bending your knee" is (extending? flexing?) your lower leg.

9 What is the name of the main muscle that
  a Flexes the upper arm?
  b Flexes the lower arm?
  c Flexes the thigh?
  d Flexes the lower leg?
  e Extends the upper arm?
  f Extends the lower arm?
  g Extends the thigh?
  h Extends the lower leg?

10 Give the approximate location of each of the following muscles and tell what movement it produces.

biceps brachii   pectoralis major
hamstrings   quadriceps femoris group
deltoid   latissimus dorsi

11 Briefly describe changes that gradually take place in bones, joints, and muscles if a person habitually gets too little exercise.

12 Explain the following terms:

atrophy   muscle tone
posture   isometric contractions
good posture   isotonic contractions
hypertrophy   tonic contractions

unit three

Systems that provide communication and control

chapter 6

The nervous system

Any group of individuals who work together on anything have to have a boss. If their many separate jobs are going to accomplish one complex task, then obviously what each worker does must be controlled or regulated. This principle holds true regardless of the nature of the complex task to be done. Suppose the task is that of giving good care to hospitalized sick people. Can you imagine what would happen with no communication between hospital workers? Utter chaos! Only through communication can their many diverse activities be controlled and made to function as a unit.

The normal body accomplishes a gigantic and enormously complex job—that of keeping itself alive and healthy. Each one of its billions of individual cells performs some function that is a part of this big function. Control of the body's billions of cells is accomplished mainly by two communication systems, namely, the nervous system and the endocrine system. Both systems transmit information from one part of the body to another, but they do it in different ways. The nervous system transmits information by means of nerve impulses conducted from one nerve cell to another. The endocrine system transmits information by means of chemicals that are secreted by ductless glands into the bloodstream and are circulated from the glands to other parts of the body. Nerve impulses and hormones communicate information to body structures—

increasing or decreasing their activities as needed for healthy survival. In other words, the communication systems of the body are also its control and integrating systems. They weld the body's hundreds of different functions into its one overall function of keeping itself alive and healthy. Our plan for this chapter is to name the organs and divisions of the nervous system, describe the coverings and fluid spaces of the brain and cord, relate basic information about the nervous system's special kinds of cells, and then discuss the various organs of the nervous system. Chapter 7 considers the autonomic nervous system and Chapter 8 discusses the endocrine system.

Organs and divisions of nervous system

The organs of the nervous system are the spinal cord, the brain, and the numerous nerves of the body. Because the brain and spinal cord occupy a midline, or central, location in the body, together they are called the central nervous system, or simply the CNS. Similarly, the usual designation for the nerves of the body is PNS, meaning the peripheral nervous system—an appropriate name because nerves extend to outlying or peripheral parts of the body. Still another division of the nervous system is the so-called autonomic nervous system (discussed in Chapter 7).

Coverings and fluid spaces of brain and spinal cord

Nervous tissue is not a sturdy tissue. Even moderate pressure can kill nerve cells, so nature safeguards the chief organs made of this tissue—the spinal cord and the brain—by surrounding them with a tough, fluid-containing membrane (the meninges) and by surrounding the membrane with bones. The *spinal meninges* form a tubelike covering around the spinal cord and line the vertebrae. Look at Fig. 6-1 and you can identify the three layers of the spinal meninges. They are the dura mater (lines the vertebrae), the pia mater (covers the cord), and the arachnoid membrane (between the dura mater and the pia mater). This middle layer of the meninges, the arachnoid membrane, resembles a cobweb with fluid filling in its spaces. The word "arachnoid" means "cobweblike." It comes from *arachne*, the Greek word for spider. Arachne is the name of the girl who was changed into a spider by Athena because she boasted of the fineness of her weaving. At least, so an ancient Greek myth tells us.

The meninges that form the protective covering around the spinal cord also extend up and around the brain to enclose it completely (Fig. 6-2). Fluid fills the arachnoid spaces of the brain meninges as well as those of the spinal cord. Also, as the next paragraph explains, there are fluid-filled spaces inside the brain called the cerebral ventricles.

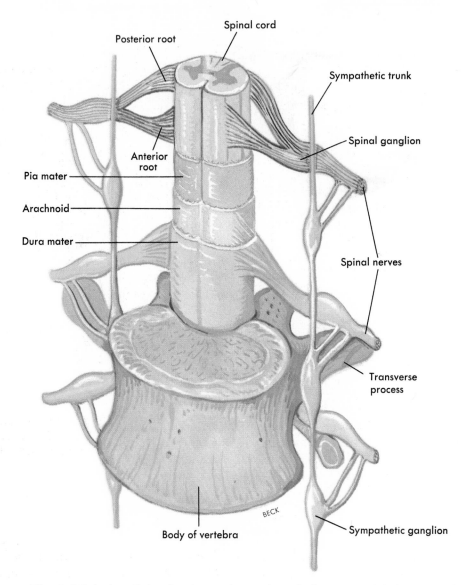

Fig. 6-1 Spinal cord showing the meninges, the spinal nerves and their roots and ganglia, and the sympathetic trunk and ganglia.

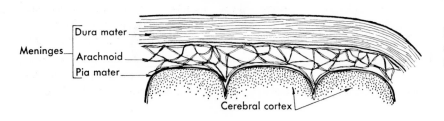

Fig. 6-2 Schematic drawing showing the structure of the meninges around the brain.

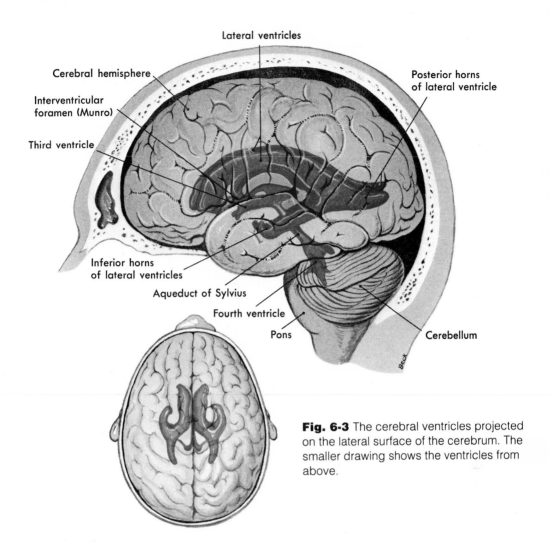

Lateral ventricles

Cerebral hemisphere

Posterior horns
of lateral ventricle

Interventricular
foramen (Munro)

Third ventricle

Inferior horns
of lateral ventricles

Aqueduct of Sylvius

Fourth ventricle

Pons

Cerebellum

Fig. 6-3 The cerebral ventricles projected on the lateral surface of the cerebrum. The smaller drawing shows the ventricles from above.

In Fig. 6-3 you can see what irregular shapes the ventricles of the brain have. This illustration can also help you visualize the location of the ventricles if you remember two things—that these large spaces lie deep inside the brain and that there are two lateral ventricles. One lies inside the right half of the cerebrum (the largest part of the human brain), and one lies inside the left half of the cerebrum.

Cerebrospinal fluid is one of the body's circulating fluids. It forms continually from fluid filtering out of the blood in a network of brain capillaries (known as the choroid plexus) and into the ventricles. From the lateral ventricles cerebrospinal fluid seeps into the third ventricle and flows down through the aqueduct of Sylvius (find this in Fig. 6-3) into the fourth ventricle. From the fourth ventricle it moves into the small tubelike central canal of the cord and out into the subarachnoid spaces. Then it moves leisurely down and around the cord and up and around the brain (in the subarachnoid spaces of their meninges) and returns to the blood (in veins of the brain).

The exact circulation route of cerebrospinal fluid may well be information you will never need to know. But remembering that this fluid forms continually from blood, circulates, and is reabsorbed into blood can be useful. It can help you understand certain abnormalities you may see. Suppose a person has a brain tumor that presses on the aqueduct of Sylvius. This blocks the way for the return of cerebrospinal fluid to the blood. Since the fluid continues forming but cannot drain away, it accumulates in the ventricles or in the meninges (subarachnoid spaces around the brain). Other conditions besides brain tumors can cause an accumulation of cerebrospinal fluid in the ventricles. An example is hydrocephalus, or "water on the brain."

Cells of nervous system

Special cells found only in the nervous system are called *neurons* (or *nerve cells*) and *neuroglia*. *Neurons* specialize in the function of transmitting impulses. *Neuroglias* are a special type of connective tissue cell. One reason for mentioning neuroglias is that one of the commonest types of brain tumor—called the *glioma*—develops from them. Neuroglias vary in size and shape. Some are relatively large cells that look somewhat like stars because of the many threadlike extensions that jut out from their surfaces. (These neuroglias are called *astrocytes*, a word that means "star cells.") Their threadlike branches attach to both neurons and small blood vessels, holding these structures close to each other. *Microglia* are smaller cells than astrocytes. Usually they remain stationary, but in inflamed or degenerating brain tissue they enlarge, move about, and act as microbe-eating scavengers. They surround the microbes, draw them into their cytoplasm, and digest them. *Phagocytosis* is the scientific name for this important cellular process.

Each neuron consists of three main parts: a main part called the *neuron cell body*, one or more branching projections called *dendrites*, and one elongated projection known as an *axon*. Identify each of these parts on each of the neurons shown in Fig. 6-4. Dendrites are the processes (projections) that transmit impulses to the neuron cell bodies, and axons are the processes that transmit impulses away from them.

Classified according to the direction in which they transmit impulses, there are three types of neurons: sensory neurons, motoneurons, and interneurons. *Sensory neurons* transmit impulses to the spinal cord and brain from all parts of the body. *Motoneurons* transmit impulses in the opposite direction—away from the brain and cord, not to them. Motoneurons do not conduct impulses to all parts of the body but only to two kinds of tissue—muscle and glandular epithelial tissue. *Interneurons* conduct impulses from sensory neurons to motoneurons. Sensory neurons are also called afferent neurons; motoneurons are called efferent neurons; interneurons are called central, or connecting, neurons.

Reflex arcs

Every moment of our lives, nerve impulses speed over neurons to and from our spinal cords and brains. If all impulse conduction ceases, life itself ceases. Only neurons can provide the rapid communication between the body's billions of cells that is necessary for maintaining life. Chemical "messages" are the only other kind of communications the body can send, and these travel much more slowly than impulses. They can move from one part of the body to another only by way of the circulating

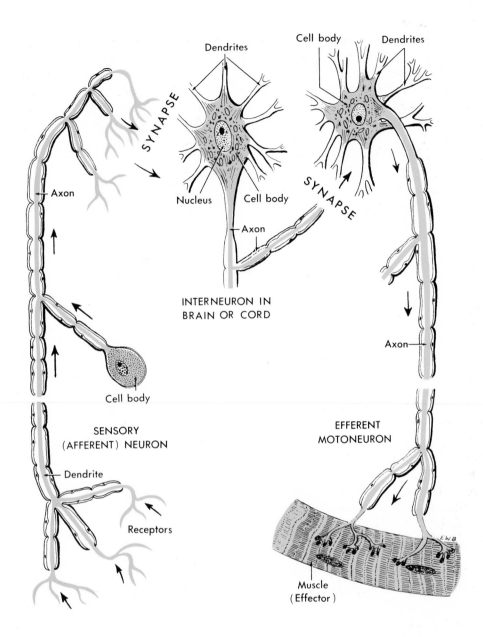

Dendrites

Cell body Dendrites

SYNAPSE

Nucleus Cell body

SYNAPSE

Axon

Axon

INTERNEURON IN
BRAIN OR CORD

Axon

Cell body

SENSORY
(AFFERENT) NEURON

EFFERENT
MOTONEURON

Dendrite

Receptors

Muscle
(Effector)

Fig. 6-4 Diagrammatic representation of structure of three kinds of neurons. Note that each neuron has three parts: a cell body and two types of extensions, dendrite(s) and axon. Arrows indicate direction of impulse conduction.

blood. Compared with impulse conduction, circulation is a very slow process indeed.

Nerve impulses can travel over literally trillions of routes—routes made up of neu-rons, because neurons are the cells that conduct impulses. Hence the routes trav-eled by nerve impulses are sometimes spo-ken of as neuron pathways. Their scientific name, however, is *reflex arcs*. The simplest kind of reflex arc is a *two-neuron arc,* so called because it consists of only two types of neurons—sensory neurons and motoneurons. *Three-neuron arcs* are the next simplest kind. They, of course, consist of all three kinds of neurons—sensory neurons, interneurons, and motoneurons. Reflex arcs are like one-way streets. They allow impulse conduction in only one direction. The next paragraph describes this direction in detail. Look frequently at Fig. 6-5 as you read it.

Impulse conduction normally starts in receptors. *Receptors* are the beginnings of dendrites of sensory neurons. None are shown in Fig. 6-5 because they are located

at some distance from the cord—in skin and mucous membranes, for example. From receptors, impulses travel the full length of the sensory neuron's dendrite, through its cell body and axon, to the branching brush-like ends of the axon. The ends of the sensory neuron's axon contact interneurons. Actually, a microscopic space separates the axon endings of one neuron from the dendrites of another neuron. This space is called a *synapse.* After crossing the synapse the impulses continue along the dendrites, cell bodies, and axons of the interneurons. Then they cross another synapse. Finally, they travel over the dendrites, cell bodies, and axons of motorneurons to a structure called an *effector* (because it "puts into ef-

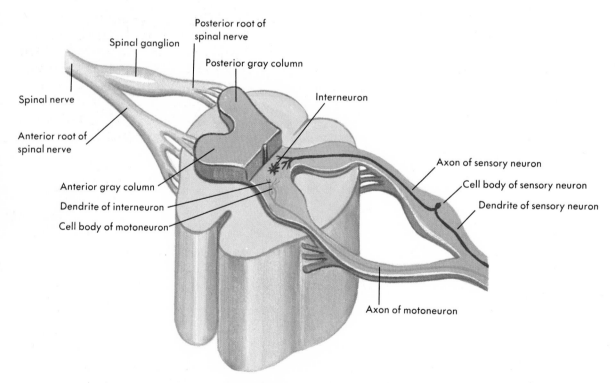

Fig. 6-5 Cross section of the spinal cord. Its interior is composed of gray matter surrounded by white matter. The left side of the diagram shows macroscopic structures only. The right side of the diagram shows locations of the dendrite, cell body, and axon (all microscopic structures) of a sensory neuron, an interneuron, and a motoneuron, the structures that compose a three-neuron reflex arc.

fect" the message brought to it by nerve impulses). Effectors are either muscles (skeletal muscles in this case, although not shown in Fig. 6-5) or glands. When impulses reach skeletal muscle cells, they cause the cells to contract. This brings us to another definition. The response to impulse conduction over reflex arcs—in this case muscle contraction—is called a *reflex*. In short, impulse conduction by a reflex arc causes a reflex to occur. Muscle contractions and gland secretion are the only two kinds of reflexes.

The term "reflex center" means the center of a reflex arc. The first part of a reflex arc consists of sensory neurons, and the last part consists of motoneurons. In spinal cord arcs the sensory neurons conduct impulses to the cord, and the motoneurons conduct them away from the cord and out to muscles. The reflex centers of all spinal cord arcs, then, lie in spinal cord gray matter. In three-neuron arcs the reflex centers consist of interneurons. In two-neuron arcs they are simply the synapses between sensory and motoneurons. We might define a reflex center as the place in a reflex arc where incoming impulses become outgoing impulses.

Fig. 6-5 reveals a number of important facts. Dendrites, cell bodies, and axons of neurons are microscopic structures that are located in macroscopic structures. For example, by comparing the right and left sides of the diagram you can see that the dendrite of the sensory neuron lies in a spinal nerve. All sensory dendrites are fairly long structures because they begin in receptors in the skin and other organs and end at neuron cell bodies in structures called ganglia located near the spinal cord or brain. The sensory neuron cell body shown in Fig. 6-5 lies in a spinal ganglion. Note that it is a small swelling on the posterior root of a spinal nerve. (Because of this location,

spinal ganglia are also called posterior root ganglia.) Each spinal ganglion contains not one sensory neuron cell body as shown in Fig. 6-5 but hundreds of them. Now turn your attention to the interneuron in this figure. All interneurons lie entirely within the gray matter of the central nervous system, that is, gray matter of either the brain or spinal cord. Gray matter forms the H-shaped inner core of the spinal cord.

Identify the motoneuron in Fig. 6-5. Observe that its dendrites and cell body are located in the cord's gray matter, specifically in the anterior horn (column) of gray matter. Not shown in the figure is the termination of this neuron's axon. It ends, as do the axons of all anterior horn motoneurons, in skeletal muscle. This fact has interesting applications. For instance, the virus responsible for poliomyelitis attacks and destroys anterior horn motoneurons. And since their axons are the only ones that conduct impulses to skeletal muscles, impulses can no longer reach the muscles supplied by the destroyed anterior horn motoneurons. With no impulses to stimulate their contraction, these muscles become paralyzed.

Nerve impulses

What are nerve impulses? Here is one widely accepted definition: a nerve impulse is a self-propagating wave of electrical negativity that travels along the surface of a neuron's cytoplasmic membrane. You might visualize this as a tiny spark sizzling its way along a fuse. Nerve impulses do not continually race along every nerve cell's surface. First they have to be initiated by a stimulus, a change in the neuron's environment. Pressure, temperature, and chemical changes are the usual stimuli. When an adequate stimulus acts on a neuron, it greatly increases the permeability of the stimulated point of its membrane to sodium ions. These positively charged ions therefore rush through the stimulated point into the interior of the neuron. The inward movement of positive ions leaves a slight excess of negative ions outside. A point of electrical negativity has been created on the neuron's surface. A nerve impulse has begun—a self-propagating wave of electrical negativity that speeds point by point along the entire length of the neuron's surface.

Neurotransmitters

Neurotransmitters are chemical compounds released in minute amounts from axon terminals into a crevice about one millionth of an inch wide, called a synaptic cleft. Almost instantly some of the neurotransmitter molecules are captured by certain protein molecules (called receptors) embedded in the cytoplasmic membrane of a neuron whose dendrites and cell body lie a mere millionth of an inch away across the synaptic cleft. Impulse conduction by this neuron is initiated by the combining of neurotransmitter with receptor protein molecules. The place where nerve impulses are transmitted from one neuron to another is called a *synapse.* The neuron that conducts impulses to a synapse is called, logically enough, a presynaptic neuron, and the one that conducts impulses away from the synapse is called a postsynaptic neuron. Now look once again at Fig. 6-5. Observe that two synapses in the reflex arc are diagrammed here—one between the sensory neuron and the interneuron and another between the interneuron and the motoneuron. At the first synapse, the sensory neuron is the presynaptic neuron and the interneuron is the postsynaptic one. At the second synapse, which neuron is presynaptic and which is postsynaptic?

A number of different compounds function as neurotransmitters. For example, the substance named acetylcholine is known to be released at some of the synapses in the spinal cord. Norepinephrine, dopamine, and serotonin are also neurotransmitters. They belong to group of compounds called catecholamines and are released at certain synapses in the brain. Chapters 7 and 8 give more information about these highly important substances.

Spinal cord
Structure

If you are of average height, your spinal cord is about 17 or 18 inches long. It lies inside the spinal column in the spinal cavity and extends from the occipital bone down to the bottom of the first lumbar vertebra. Place your hands on your hips, and they will line up with your fourth lumbar vertebra. Your spinal cord ends just above this level. The spinal meninges, however, do not end here. They continue down almost to the end of the spinal column—a useful fact for a physician to know. It means that he can do a lumbar puncture to remove spinal fluid with no fear of damaging the cord, provided that he inserts the needle into the meninges below the level where the cord ends.

Look now at Fig. 6-5. Notice the H-shaped core of the spinal cord. It consists of gray matter and so is composed mainly of dendrites and cell bodies of neurons. Columns of white matter form the outer portion of the cord and bundles of nerve fibers—the spinal tracts—make up the white columns.

Functions

To try to understand spinal cord functions, let us start by thinking about a hotel telephone switchboard. Suppose a guest in Room 108 calls the switchboard operator and asks for Room 520, and in a second or so someone in that room answers. Very briefly, three events took place: a message traveled into the switchboard, a connection was made in the switchboard, and a message traveled out from the switchboard. The telephone switchboard provided the connection that made possible the completion of this call. Or we might say that it transferred the incoming call to an outgoing line. The spinal cord functions similarly. It contains the centers for thousands and thousands of reflex arcs. Look back at Fig. 6-5. The interneuron shown there is an example of a spinal cord reflex center. It switches, or transfers, incoming sensory impulses to outgoing motor impulses, thereby making it possible for a reflex to occur. Reflexes that result from conduction over arcs whose centers lie in the cord are called spinal cord reflexes. Two common kinds of spinal cord reflexes are the withdrawal reflexes and the jerk reflexes. An example of a withdrawal reflex is pulling one's hand away from a hot surface. The familiar knee jerk is an example of a jerk reflex.

In addition to functioning as the primary reflex center of the body, the spinal cord also functions as a conductor to and from the brain. It contains bundles of nerve fibers that are called spinal tracts. Some of these conduct sensory impulses up to the brain, others conduct motor impulses down from the brain. Therefore, if an injury cuts the cord all the way across, impulses can no longer travel to the brain from any part of the body located below the injury, nor can they travel from the brain down to these parts. In short, this kind of spinal cord injury produces both a loss of sensation (anesthesia) and a loss of the ability to make voluntary movements (paralysis).

Spinal nerves
Structure

Thirty-one pairs of nerves attach to the spinal cord in the following order: eight pairs attach to the cervical segments, twelve pairs attach to the thoracic segments, five pairs attach to the lumbar segments, five pairs attach to the sacrospinal segments, and one pair attaches to the coccygeal segment. Unlike cranial nerves, spinal nerves have no special names; instead, a letter and number identify each one. C1, for example, indicates the pair of spinal nerves attached to the first segment of the cervical part of the cord, T8 indicates those attached to the eighth segment of the thoracic part of the cord, and so on.

Function

Spinal nerves conduct impulses between the spinal cord and the parts of the body not supplied by cranial nerves. The spinal nerve shown in Fig. 6-5 contains, as do all spinal nerves, both sensory and motor fibers. Spinal nerves therefore function to make possible both sensations and movements.

A disease or injury that prevents conduction by a spinal nerve therefore results in both a loss of feeling and a loss of movement in the part supplied by that nerve. You may have known someone who suffered with sciatica. This is an inflammation of the spinal nerve branch called the sciatic nerve—the largest nerve in the body, incidentally. Shingles is another painful and fairly common spinal nerve inflammation.

Divisions of the brain

The brain, one of our largest organs, consists of the following major divisions named in ascending order beginning with the lowest part: brain stem (medulla, pons, and

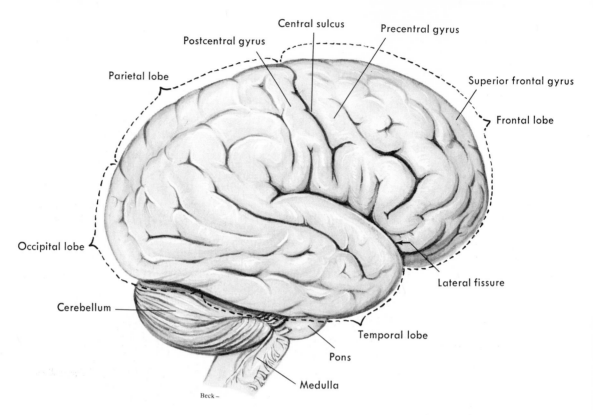

Fig. 6-6 Lateral view of the brain.

midbrain), cerebellum, diencephalon (hypothalamus and thalamus), and cerebrum. Observe in Fig. 6-6 the location and relative sizes of the medulla, pons, cerebellum, and cerebrum. Identify the midbrain in Fig. 6-7.

Brain stem

The lowest part of the brain stem is formed by the medulla. Immediately above the medulla lies the pons and above that the midbrain. Together, these three structures are called the brain stem. Look at Fig. 6-7 and you can see why.

The *medulla* is a bulb-shaped extension of the spinal cord. It lies just inside the cranial cavity above the large hole in the occipital bone called the foramen magnum. Like the spinal cord, the medulla consists of both gray and white matter. But their arrangement differs in the two organs. In the medulla, bits of gray matter mix closely and intricately with white matter to form what is called the reticular formation (reticular means netlike). In the cord, gray and white matter do not intermingle; gray matter forms the interior core of the cord and white matter surrounds it. The *pons* and *midbrain*, like the medulla, consist of white matter and scattered bits of gray matter.

All three parts of the brain stem function as two-way conduction paths. Sensory fibers conduct impulses up from the cord to

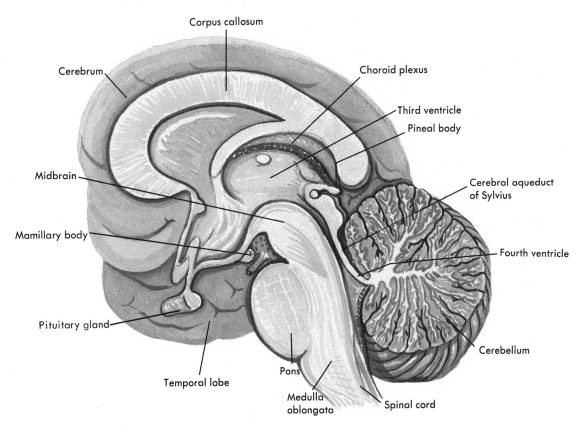

Corpus callosum

Cerebrum

Choroid plexus

Third ventricle

Pineal body

Midbrain

Cerebral aqueduct
of Sylvius

Mamillary body

Fourth ventricle

Pituitary gland

Cerebellum

Temporal lobe

Pons

Medulla
oblongata

Spinal cord

Fig. 6-7 Sagittal section of the brain. (Note position of midbrain.)

other parts of the brain and motor fibers conduct impulses down from the brain to the cord. In addition, many important reflex centers lie in the brain stem. The cardiac, respiratory, and vasomotor centers (collectively called the vital centers), for example, are located in the medulla. Impulses from these centers control the heartbeat, respirations, and blood vessel diameter.

Diencephalon

The diencephalon is the small but important part of the brain located between the midbrain below and the cerebrum above. It consists of two major structures, namely, the hypothalamus and the thalamus.

The *hypothalamus,* as its name suggests, is located below the thalamus. Major parts of the hypothalamus are the posterior pituitary gland, the stalk that attaches it to the undersurface of the brain, and two pairs of clusters of neuron cell bodies (named the paraventricular and supraoptic nuclei) located in the side walls of the third ventricle. Identify the pituitary gland, its stalk, and the third ventricle in Fig. 6-7. The mammillary body shown in this figure is also part of the hypothalamus.

The old adage, "Don't judge by appearances," applies well to appraising the importance of the hypothalamus. Measured by size, it is one of the least significant parts of the brain, but measured by its contribution to healthy survival, the hypothalamus is one of the most important of all brain structures. Impulses from neurons whose dendrites and cell bodies lie in the hypothalamus are conducted by their axons to neurons located in the spinal cord, and many of them are then relayed to muscles and glands all over the body. Thus the hypothalamus exerts a major control over virtually all internal organs. Among the vital functions it helps control are the beating of the heart, the constriction and dilatation of blood vessels, and the contractions of the stomach and intestines.

Neurons in the supraoptic and paraventricular nuclei of the hypothalamus function in a surprising way. They make the hormones that the posterior pituitary gland secretes into the blood. Because one of these hormones (called antidiuretic hormone or ADH) affects the volume of urine excreted, the hypothalamus plays an essential role in maintaining the body's water balance.

Some of the neurons in the hypothalamus function as endocrine (ductless) glands. Their axons secrete chemicals, called releasing hormones, into the blood, which then carries them to the anterior pituitary gland. Releasing hormones, as their name suggests, control the release of certain anterior pituitary hormones. These in turn influence hormone secretion by various other endocrine glands. Thus the hypothalamus indirectly helps control the functioning of every cell in the body.

The hypothalamus performs several other important functions. It helps regulate our appetites and therefore the amount of food we eat. It also plays a central role in maintaining normal body temperature.

Located deep inside each half of the cerebrum, lateral to the third ventricle, lies a rounded mass of gray matter. This is the *thalamus,* the other major part of the dien-

cephalon. Each thalamus consists chiefly of dendrites and cell bodies of neurons whose axons extend to various sensory areas of the cerebral cortex.

The thalamus performs the following functions.

1 It helps produce sensations. Its neurons relay impulses to the cerebral cortex from the various sense organs of the body (except possibly those responsible for the sense of smell).

2 It associates sensations with emotions. Almost all sensations are accompanied by a feeling of some degree of pleasantness or unpleasantness. Just how these pleasant and unpleasant feelings are produced is not known except that they seem to be associated with the arrival of sensory impulses in the thalamus.

3 It plays a part in the so-called arousal or alerting mechanism.

Cerebrum

The cerebrum is the largest and uppermost part of the brain. If you were to look at the outer surface of the cerebrum, perhaps the first features you would notice are its many ridges and grooves. The ridges are called convolutions or gyri and the grooves are called fissures. The deepest fissure, namely the longitudinal fissure, divides the cerebrum into right and left halves or hemispheres. They are almost separate structures except for their lower midportions, which are connected by a structure called the corpus callosum (Fig. 6-7). Shallower fissures subdivide each cerebral hemisphere into four major lobes and each lobe into numerous convolutions. The lobes are named for the bones that lie over them: the frontal lobe, the parietal lobe, the temporal lobe, and the occipital lobe. Identify these in Figs. 6-6 and 6-8.

A thin layer of gray matter, made up of neuron dendrites and cell bodies, composes the surface of the cerebrum. Its name is the cerebral cortex. White matter, made up of bundles of nerve fibers (tracts), composes most of the interior of the cerebrum. Within this white matter, however, are a few islands of gray matter known as the basal ganglia, whose functioning is essential for producing our automatic movements and postures. Parkinson's disease is a disease of the basal ganglia. Because shaking or tremors are common symptoms of Parkinson's disease, it is also called "shaking palsy."

What functions does the cerebrum perform? This is a hard question to answer briefly, because the neurons of the cerebrum do not function alone. They function with many other neurons located in many other parts of the brain and in the spinal cord. Neurons of these structures are continually bringing impulses to cerebral neurons and continually transmitting impulses away from them. If all other neurons were functioning normally and only cerebral neurons were not functioning, here are some of the things that you could not do. You could not think or use your will. You could not decide to make the smallest movement nor could you make it. You could not see or hear. You could not experience any of the sensations that make life so rich and varied. Nothing would anger you or frighten you, and nothing would bring you joy or sorrow. You would, in short, be unconscious. These five terms, then, sum up cerebral functions: consciousness, mental processes, sensations, emotions, and willed movements. Fig. 6-8 shows the areas of the cerebral cortex essential for willed movements, general sensations, vision, hearing, and normal speech.

Injury or disease can destroy neurons. An all too common example is the destruction of neurons of the motor area of the cerebrum that results from a cerebrovascular accident (CVA)—medical parlance for a hemorrhage from cerebral blood vessels. When this happens, the victim can no longer voluntarily move the parts of his body on the side opposite to the side on which the cerebral hemorrhage occurred. In nontechnical language, he has suffered a stroke. Note in Fig. 6-8 the location of the motor area in the frontal lobe of the cerebrum.

Cerebellum

Structure. Look at Figs. 6-3, 6-6, 6-7, and 6-9 to find the location, appearance, and size of the cerebellum. The cerebellum is the second largest part of the human brain. It lies under the occipital lobe of the cerebrum. Another fact about the cerebellum, one visible in Fig. 6-7, is that gray matter composes its outer layer and white matter composes the bulk of its interior.

Function. Most of our knowledge about cerebellar functions has come from observing patients who have some sort of disease of the cerebellum and animals who have had the cerebellum removed. From such observations we know that the cerebellum plays an essential part in the production of normal movements. Perhaps a few examples will make this clear. A patient who has a tumor of the cerebellum frequently loses his balance and topples over; he may reel like a drunken man when he walks. He probably cannot coordinate his muscles normally. He may complain, for instance, that he is clumsy about everything he does—that he cannot even drive a nail or draw a straight line. With the loss of normal cerebellar functioning, he has lost the ability to make precise movements. The general functions of the cerebellum, then, are to produce smooth coordinated movements, maintain equilibrium, and sustain normal postures.

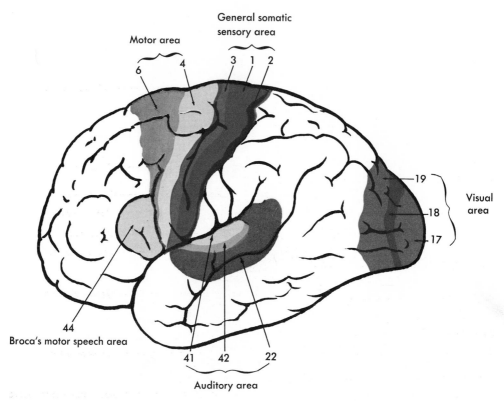

Fig. 6-8 Localization of function in the cerebral cortex.

Cranial nerves

Twelve pairs of cranial nerves attach to the undersurface of the brain, mostly from the brain stem. Fig. 6-9 shows the attachments of a few of these nerves. Their fibers conduct impulses between the brain and various structures in the head and neck and in the thoracic and abdominal cavities. For instance, the second cranial nerve (optic nerve) conducts impulses from the eye to the brain, where these impulses produce the sensation of vision. The third cranial nerve (oculomotor nerve) conducts impulses from the brain to certain muscles of the eye, where they cause contractions that move the eye. The tenth cranial nerve (vagus nerve) conducts impulses between the medulla and various structures in the neck and thoracic and abdominal cavities. The names of each of the cranial nerves and a brief description of their functions are listed in Table 6-1.

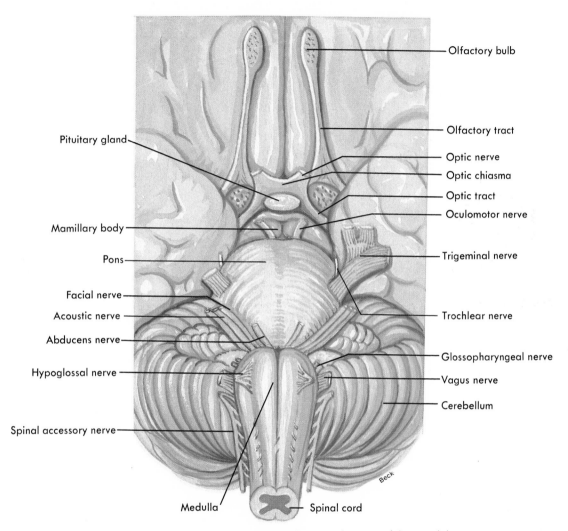

Olfactory bulb

Olfactory tract

Optic nerve

Optic chiasma

Optic tract

Oculomotor nerve

Trigeminal nerve

Trochlear nerve

Glossopharyngeal nerve

Vagus nerve

Cerebellum

Pituitary gland

Mamillary body

Pons

Facial nerve

Acoustic nerve

Abducens nerve

Hypoglossal nerve

Spinal accessory nerve

Medulla

Spinal cord

Beck

Fig. 6-9 Undersurface of the brain showing attachments of the cranial nerves.

Table 6-1 Cranial nerves

Nerve*	Conducts impulses	Functions
I Olfactory	From nose to brain	Sense of smell
II Optic	From eye to brain	Vision
III Oculomotor	From brain to eye muscles	Eye movements
IV Trochlear	From brain to external eye muscles	Eye movements
V Trigeminal (or trifacial)	From skin and mucous membrane of head and from teeth to brain; also from brain to chewing muscles	Sensations of face, scalp, and teeth; chewing movements
VI Abducens	From brain to external eye muscles	Turning eyes outward
VII Facial	From taste buds of tongue to brain; from brain to face muscles	Sense of taste; contraction of muscles of facial expression
VIII Acoustic	From ear to brain	Hearing; sense of balance
IX Glossopharyngeal	From throat and taste buds of tongue to brain; also from brain to throat muscles and salivary glands	Sensations of throat, taste, swallowing movements; secretion of saliva
X Vagus	From throat, larynx, and organs in thoracic and abdominal cavities to brain; also from brain to muscles of throat and to organs in thoracic and abdominal cavities	Sensations of throat, larynx, and of thoracic and abdominal organs; swallowing, voice production, slowing of heartbeat, acceleration of peristalsis
XI Spinal accessory	From brain to certain shoulder and neck muscles	Shoulder movements; turning movements of head
XII Hypoglossal	From brain to muscles of tongue	Tongue movements

*The first letter of the words of the following sentence are the first letters of the names of cranial nerves: "On Old Olympus' Tiny Tops A Finn and German Viewed Some Hops." Many generations of students have used this or a similar sentence to help them remember the names of cranial nerves.

Sense organs

If you were asked to name the sense organs, what organs would you name? Can you think of any besides the eyes, ears, nose, and taste buds? Acually there are millions of other sense organs—all receptors are microscopic-sized sense organs. Receptors, you will recall, are the beginnings of dendrites of sensory neurons.

Receptors are generously scattered about in almost every part of the body. To demonstrate this fact, try pricking any point of

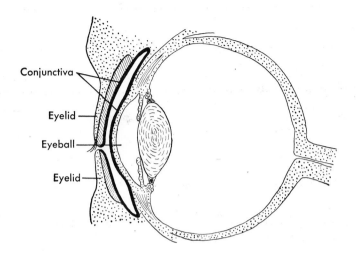

Fig. 6-10 Cross-sectional diagram of the eye showing location of the conjunctiva.

Labels: Conjunctiva, Eyelid, Eyeball, Eyelid

your skin with a fine needle. You can hardly miss stimulating at least one receptor and almost instaneously experiencing a sensation of mild pain. Stimulation of some receptors leads to the sensation of heat. Stimulation of other receptors gives the sensation of cold, and stimulation of still others gives the sensation of touch or pressure. When special receptors in the muscles and joints are stimulated, you sense the position of the different parts of the body and know whether they are moving and in which direction they are moving without even looking at them. Perhaps you have never realized that you have this sense of position and movement—a sense called *proprioception* or *kinesthesia*. Let us turn our attention now to two complex and remarkable sense organs—the eyes and ears.

Eye

When you look at a person's eye, you see only a small part of the whole eye. Three layers of tissue form the eyeball: the sclera, the choroid, and the retina. The outer layer of *sclera* consists of tough fibrous tissue. What we call the "white" of the eye is part of the front surface of the sclera. The other part of the front surface of the sclera is

called the cornea and is sometimes spoken of as the "window" of the eye because of its transparency. At a casual glance, however, it does not look transparent but appears blue or brown or gray or green because it lies over the *iris*, the colored part of the eye. Mucous membrane known as the *conjunctiva* covers the entire front surface of the eyeball and lines both the upper and lower lids (Fig. 6-10).

Two involuntary muscles make up the front part of the middle, or *choroid* coat, of the eyeball. One is the *iris*, the colored structure seen through the cornea, and the other is the *ciliary muscle* (part of the ciliary body shown in Fig. 6-11). What appears to be a black center in the iris is really a hole in this doughnut-shaped muscle; it is the *pupil* of the eye. Some of the fibers of the iris are arranged like spokes in a wheel. When they contract the pupils dilate, letting in more light rays. Other fibers are circular. When they contract the pupils constrict, letting in fewer light rays. Normally, the pupils constrict in bright light and dilate in dim light.

The *lens* of the eye lies directly behind the pupil. It is held in place by a ligament attached to the ciliary muscle. When we look

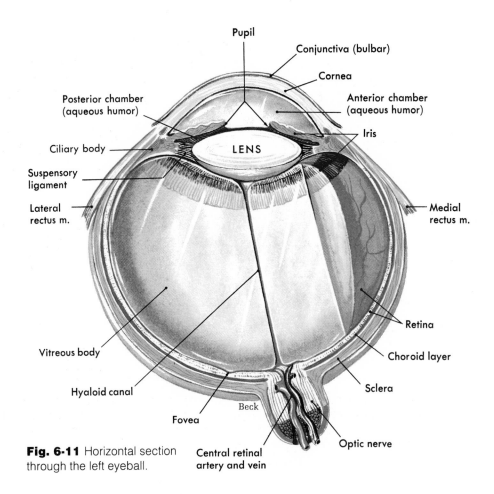

Pupil

Conjunctiva (bulbar)

Cornea

Posterior chamber
(aqueous humor)

Anterior chamber
(aqueous humor)

Iris

Ciliary body

LENS

Suspensory
ligament

Lateral
rectus m.

Medial
rectus m.

Retina

Choroid layer

Vitreous body

Sclera

Hyaloid canal

Beck

Fovea

Optic nerve

Fig. 6-11 Horizontal section through the left eyeball.

Central retinal
artery and vein

at distant objects, the ciliary muscle is relaxed, and the lens has only a slightly curved shape. To focus on near objects, however, the ciliary muscle must contract. As it contracts, it pulls the choroid coat forward toward the lens thus causing the lens to become more bulging and more curved. Most of us become more farsighted as we grow older. The reason we do is that our lenses lose their elasticity and can no longer bulge enough to bring near objects into focus. Presbyopia, or "oldsightedness," is the name for this condition.

The *retina,* or innermost coat of the eye-ball, contains microscopic structures called rods and cones because of their shapes. Both are receptors for vision. Dim light can stimulate the rods, but fairly bright light is necessary to stimulate the cones. In other words, *rods* are the receptors for night vision and *cones* for daytime vision. (Cones are also the receptors for color vision.)

Fluids fill the hollow inside of the eyeball. They maintain the normal shape of the eyeball and help refract light rays; that is, the fluids bend light rays so as to bring them to a focus on the retina. *Aqueous humor* is the name of the fluid in front of the lens (in the

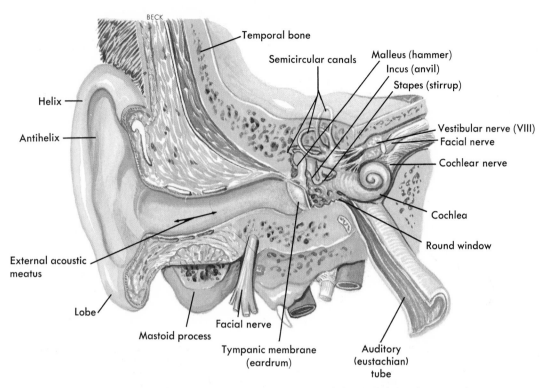

Fig. 6-12 Components of the ear. External ear consists of auricle (pinna), external acoustic meatus (ear canal), and tympanic membrane (eardrum). Middle ear (tympanic cavity) includes malleus (hammer), incus (anvil), and stapes (stirrup). Internal ear contains semicircular canals, vestibule, and cochlea.

anterior cavity of the eye) and *vitreous humor* is the name of the jellylike fluid behind the lens (in the posterior cavity).

Ear

The ear is much more than a mere appendage on the side of the head. A large part of the ear—and by far its most important part—lies hidden from view deep inside the temporal bone. Part of the external ear, all of the middle ear, and all of the internal ear are located here.

The *external ear* has two parts: the pinna (or auricle) and the ear canal (or external acoustic meatus). The *pinna* is the appendage on the side of the head. The *ear canal* is a curving tube in the temporal bone that leads from the pinna to the middle ear.

The *middle ear* (or tympanic cavity) is a tiny cavity hollowed out of the temporal bone. This cavity is lined with mucous membrane and contains three very small bones. The names of these ear bones (Fig. 6-12) are Latin words that describe their shapes—*malleus* (hammer), *incus* (anvil), and *stapes* (stirrup). The *tympanic membrane* (commonly called the eardrum) separates the middle ear from the external ear canal. The "handle" of the malleus attaches to the inside of the tympanic membrane, and the "head" attaches to the incus. The incus attaches to the stapes, and the stapes fits into a small opening, the *oval window*, that opens into the internal ear. A point worth mentioning, because it explains the frequent spread of infection from the throat to the ear, is the fact that a tube—the *auditory* or *eustachian tube*—connects the throat with the middle ear. The mucous lining of the middle ears, eustachian tubes, and throat are extensions of one continuous membrane. Consequently, a sore throat may spread to produce a middle ear infection *(otitis media)*. It can even cause a mastoid infection *(mastoiditis)*. Mastoid spaces (sinuses) also open into the middle ear cavity. Because the mucous lining of the mastoid sinuses is also continuous with the mucous lining of the middle ear, it provides a direct route for a middle-ear infection to spread to produce *mastoiditis*.

The *internal ear* is made of bone and a membrane inside the bone. Because of its complicated shape, the internal ear is called a *labyrinth*. It has three parts: *vestibule, semicircular canals*, and *cochlea* (Fig. 6-12). The semicircular canals are three half circles; the cochlea is shaped like a snail shell, which is what the word "cochlea" means.

Two special sense organs for two different kinds of sensations—hearing and balance—are located in the internal ear. The hearing sense organ, which lies inside the cochlea, is called the *organ of Corti*. There are two balance, or equilibrium, sense organs—one called the *macula*, which is located in the vestibule of the inner ear, and the other called the *crista*. The crista is located in the semicircular canals.

Outline summary

Organs and divisions of nervous system

A Central nervous system (CNS)—brain and spinal cord.

B Peripheral nervous system (PNS)—all nerves

Coverings and fluid spaces of brain and spinal cord

A Coverings
  1 Cranial bones and vertebrae
  2 Cerebral and spinal meninges
B Fluid spaces—subarachnoid spaces of meninges, central canal inside cord, and ventricles in brain

Cells of nervous system

A Neuroglias—connective tissue cells of two main types
  1 Astrocytes—star-shaped cells that anchor small blood vessels to neurons
  2 Microglias—small cells that move about in inflamed brain tissue carrying on phagocytosis: that is, they engulf and destroy microorganisms and other injurious particles
B Neurons or nerve cells
  1 Consist of three main parts: dendrites—conduct impulses to cell body of neuron; cell body of neuron; and axon—conducts impulses away from cell body of neuron
  2 Neurons classified according to function as sensory—conduct impulses to spinal cord and brain; motoneurons—conduct impulses away from brain and cord out to muscles and glands; and interneurons—conduct impulses from sensory neurons to motoneurons

Reflex arcs

A Nerve impulses are conducted from receptors to effectors over neuron pathways or reflex arcs; conduction by reflex arc results in a reflex, that is, contraction by a muscle or secretion by a gland.
B Simplest reflex arcs called two-neuron arcs—consist of sensory neurons synapsing in spinal cord with motoneurons; three-neuron arcs consist of sensory neurons synapsing in spinal cord with interneurons that synapse with motoneurons

Nerve impulses

A Definition—self-propagating wave of electrical negativity that travels along surface of neuron membrane
B Mechanism
  1 Stimulus increases permeability of neuron membrane to positive sodium ions
  2 Inward movement of positive sodium ions

leaves slight excess of negative ions outside at stimulated point; marks beginning of nerve impulses

Neurotransmitters

A Definition—chemical compounds released from axon terminals (of presynaptic neuron) into synaptic cleft

B Neurotransmitters bind to specific receptor molecules in membrane of postsynaptic neuron, thereby stimulating impulse conduction by it

C Names of neurotransmitters—acetylcholine, catecholamines (norepinephrine, dopamine, serotonin), and other compounds

Spinal cord

A Outer part composed of white matter made up of many bundles of axons called tracts; interior composed of gray matter made up mainly of neuron dendrites and cell bodies

B Functions as center for all spinal cord reflexes; sensory tracts conduct impulses to brain and motor tracts conduct impulses from brain

Spinal nerves

A Structure—contain dendrites of sensory neurons and axons of motoneurons

B Functions—conduct impulses necessary for sensations and voluntary movements

Divisions of the brain

A Brain stem
  1 Consists of three parts of brain; named in ascending order, they are the medulla, pons, and midbrain
  2 Structure—medulla, pons, and midbrain; consist of white matter with bits of gray matter scattered through them
  3 Function—gray matter in brain stem functions as reflex centers, for example, for heartbeat, respirations, and blood vessel diameter; sensory tracts in brain stem conduct impulses to higher parts of brain; motor tracts conduct from higher parts of brain to cord

B Diencephalon
  1 Structure and function of hypothalmus
    a Consists mainly of posterior pituitary gland, pituitary stalk, and the paraventricular and supraoptic nuclei
    b Acts as the major center for controlling autonomic nervous system; therefore helps control the functioning of most internal organs
    c Controls hormone secretion by both anterior and posterior pituitary glands; therefore indirectly helps control hormone secretion by most other endocrine glands
    d Acts as center for controlling appetite; therefore helps regulate amount of food eaten and body weight
    e Functions in some way to maintain the waking state
    f Probably contains reward and punishment centers
  2 Structure and function of thalamus
    a Rounded mass of gray matter in each cerebral hemisphere; located lateral to each side of third ventricle
    b Relays sensory impulses to cerebral cortex sensory areas
    c Functions in some way to produce emotions of pleasantness or unpleasantness associated with sensations

C Cerebrum
  1 Largest part of human brain
  2 Outer layer of gray matter called cerebral cortex; made up of lobes, which are made up of convolutions; cortex composed mainly of neuron dendrite and cell bodies
  3 Interior of cerebrum composed mainly of nerve fibers arranged in bundles called tracts
  4 Functions of cerebrum—mental processes of all types, including sensations, consciousness, and voluntary control of movements

D Cerebellum
  1 Second largest part of human brain
  2 Helps control muscle contractions so that they produce coordinated movements so we can maintain balance, move smoothly, and sustain normal postures

Cranial nerves

(See Table 6-1)

Sense organs

A All receptors (beginning of dendrites of sensory neurons) are sense organs

B Special sense organs—eyes and ears

C Kinds of senses—many more than the familiar five senses: for example, several kinds of touch senses; proprioception—sense of position and movement.

D Eye
  1 Coats of eyeball
    a Sclera—tough outer coat; whites of eye; cornea is transparent part of sclera over iris
    b Choroid—front part of this coat made up of ciliary muscle and iris, the colored part of eye; pupil is hole in center of iris; contraction of iris muscle dilates or constricts pupil
    c Retina—innermost coat of eye; contains rods (receptors for night vision) and cones (receptors for day vision and color vision)
  2 Conjunctiva—mucous membrane that covers front surface of eyeball and lines lid
  3 Lens—transparent body behind pupil; focuses light rays on retina
  4 Eye fluids
    a Aqueous humor—in anterior cavity in front of lens
    b Vitreous humor—in posterior cavity behind lens

E Ear
  1 External ear—consists of pinna and auditory canal
  2 Middle ear—contains auditory bones (malleus, incus, and stapes); lined with mucous membrane; eustachian tubes and mastoid sinuses open into middle ear

  3 Internal ear, or labyrinth—vestibule, semicircular canals, and cochlea are three divisions; organ of Corti, sense organ of hearing, lies in cochlea; sense organs of equilibrium are the macula and the crista; macula lies in vestibule and crista lies in semicircular canals

New words

anesthesia	presynaptic neuron
axon	postsynaptic neuron
dendrite	receptor molecules
effectors	receptors
neuroglia	reflex
neurons	reflex arc
paralysis	synapse
phagocytosis	synaptic cleft
	tracts

Review questions

1 What general function does the nervous system perform?
2 What other system performs the same general function as the nervous system?
3 What general functions does the spinal cord perform?
4 What does "CNS" mean? "PNS"?
5 What are the meninges?
6 Why is the medulla considered the most vital part of the brain?
7 What general functions does the cerebellum perform?
8 What general functions does the cerebrum perform?

9 What general functions do spinal nerves perform?

10 What are some of the functions performed by cranial nerves?

11 Which pair or pairs of cranial nerves would you nickname "seeing nerves," "hearing nerves," "smelling nerves," and "tasting nerves?"

12 Would a person be blind, deaf, or neither, if both of his eighth cranial nerves atrophied?

13 Describe as fully as you can the structure of the eye.

14 Describe as fully as you can the structure of the ear.

15 What is each of the following?

conjuctiva   organ of Corti
cornea   retina
iris

16 Explain briefly why most old people are farsighted.

17 Define briefly each of the terms listed under "New words."

18 Identify each of the following:

interneuron   somatic motoneuron
motoneuron   spinal ganglion
reflex center   synapse
sensory neuron   visceral motoneuron

chapter 7

The autonomic nervous system

The titles of this chapter and the preceding one are somewhat misleading. They may make you think that the body has two nervous systems. This is not so. It has only one nervous system, which consists of two major subdivisions called the somatic nervous system and the autonomic nervous system. Many structures make up these two divisions, but they function together as a single unit.

Definitions

If you are to understand the material presented in this chapter, you will need to know the meanings of the following terms.

1 The *autonomic nervous system*, according to one definition, consists of certain motoneurons, that is, the ones that conduct impulses from the spinal cord or brain stem to three kinds of tissues, namely, cardiac muscle tissue, smooth muscle tissue, and glandular epithelial tissue. According to another definition, the autonomic nervous system consists of the parts of the nervous system that regulate the body's automatic or involuntary functions, for example, the beating of the heart, the contracting of the stomach and intestines, and the secreting of chemical compounds by glands.

2 *Autonomic neurons* are the motoneurons that make up the autonomic nervous system. The dendrites and cell bodies of some autonomic neurons are located in gray matter of the spinal cord or brain stem. Their axons extend out from these structures and terminate in ganglia. These autonomic neurons are called *preganglionic neurons* because they conduct impulses before they reach a ganglion. In the ganglia the axon endings of preganglionic neurons synapse with the dendrites or cell bodies of postganglionic neurons. *Postganglionic neurons*, as their name suggests, conduct impulses after they reach a ganglion, that is, they conduct from the ganglion to cardiac muscle, or smooth muscle, or glandular epithelial tissue.

3 *Autonomic or visceral effectors* are the tissues to which autonomic neurons conduct impulses. Specifically, visceral effectors are cardiac muscle that makes up the wall of the heart, smooth muscle that partially makes up the walls of blood vessels and other hollow internal organs, and glandular epithelial tissue that makes up the secreting part of glands. Do you recall what kind of tissue makes up somatic effectors, the structures to which somatic motoneurons conduct impulses? If not, please reread p. 87.

Names of divisions

The autonomic nervous system consists of two divisions called the sympathetic system and the parasympathetic system. Both of these divisions, in turn, consist of various autonomic ganglia and nerves. Another name for the sympathetic system is the thoracolumbar system, an appropriate name because sympathetic ganglia are connected to the thoracic and lumbar regions of the spinal cord. Parasympathetic ganglia are connected to the brain stem and to the sacral segments of the spinal cord. Another name for the parasympathetic system therefore is the craniosacral system.

Sympathetic nervous system
Structure

Sympathetic preganglionic neurons have their dendrites and cell bodies in gray matter of the thoracic and upper lumbar segments of the spinal cord. Look now at the right side of Fig. 7-1. Follow the course of the axon of the sympathetic preganglionic neuron shown there. Note that it leaves the cord in an anterior root of a spinal nerve. It next enters the spinal nerve but soon leaves it to extend to and through a sympathetic ganglion and terminate in a collateral ganglion. Here it synapses with several postganglionic neurons whose axons extend out to terminate in visceral effectors. Although not shown in Fig. 7-1, branches of the preganglionic axon ascend and descend to terminate in ganglia above and below their point of origin. All sympathetic preganglionic axons therefore synapse with many postganglionic neurons, and these frequently terminate in widely separated organs. Hence sympathetic responses are usually widespread, involving many organs and not just one.

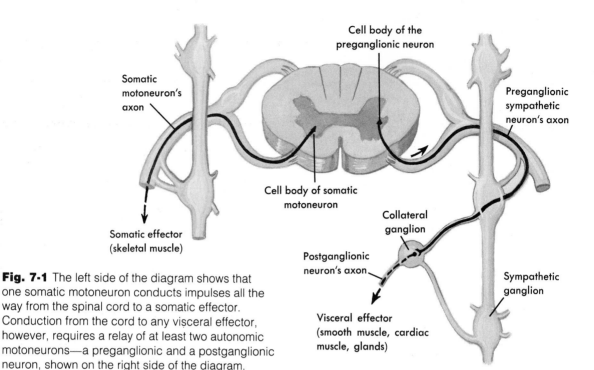

Fig. 7-1 The left side of the diagram shows that one somatic motoneuron conducts impulses all the way from the spinal cord to a somatic effector. Conduction from the cord to any visceral effector, however, requires a relay of at least two autonomic motoneurons—a preganglionic and a postganglionic neuron, shown on the right side of the diagram.

Sympathetic postganglionic neurons have their dendrites and cell bodies in sympathetic ganglia or in collateral ganglia. Sympathetic ganglia are located in front of and at each side of the spinal column. Because short fibers extend between the sympathetic ganglia, they look a little like two chains of beads and are often referred to as the "sympathetic chain ganglia." Axons of sympathetic postganglionic neurons travel in spinal nerves to blood vessels, sweat glands, and arrector hair muscles all over the body. Separate autonomic nerves distribute many sympathetic postganglionic axons to various internal organs.

Functions

The sympathetic nervous system functions as an emergency system. Impulses over sympathetic fibers take control of many of our internal organs when we exercise strenuously and when strong emotions—anger, fear, hate, anxiety—buffet us. In short, whenever we must cope with stress of any kind, sympathetic impulses increase to many visceral effectors and rapidly produce widespread changes within our bodies. The righthand column of Table 7-1 indicates many of these sympathetic responses. The heart beats faster. Most blood vessels constrict, causing blood pressure to shoot up. Blood vessels in skeletal muscles dilate, supplying the muscles with more blood. Sweat glands and adrenal glands secrete more abundantly. Salivary and other digestive glands secrete more sparingly. Digestive tract contractions (peristalsis) become sluggish, hampering digestion. To-

Table 7-1 Autonomic functions

Visceral effectors	Parasympathetic control	Sympathetic control
Heart muscle	Slows heartbeat	Accelerates heartbeat
Smooth muscle		
Of most blood vessels	None	Constricts blood vessels
Of blood vessels in skeletal muscles	None	Dilates blood vessels
Of digestive tract	Increases peristalsis	Decreases peristalsis; inhibits defecation
Of anal sphincter	Inhibits — opens sphincter for defecation	Stimulates — closes sphincter
Of urinary bladder	Stimulates — contracts bladder	Inhibits — relaxes bladder
Of urinary sphincters	Inhibits — opens sphincter for urination	Stimulates — closes sphincter
Of eye Iris	Stimulates circular fibers — constriction of pupil	Stimulates radial fibers — dilation of pupil
Ciliary	Stimulates — accommodation for near vision (bulging of lens)	Inhibits — accommodation for far vision (flattening of lens)
Of hairs (pilomotor muscles)	No parasympathetic fibers	Stimulates — "goose pimples"
Glands		
Adrenal medulla	None	Increases epinephrine secretion
Sweat glands	None	Increases sweat secretion
Digestive glands	Increases secretion of digestive juices	Decreases secretion of digestive juices

gether these sympathetic responses make us ready for strenuous muscular work, or, as the famous physiologist, Walter Cannon, vividly stated, they prepare us for "fight or flight." The group of changes induced by sympathetic control is known as the "fight-or-flight syndrome."

Parasympathetic nervous system

Structure

The dendrites and cell bodies of parasympathetic preganglionic neurons are located in gray matter of the brain stem and the sacral segments of the spinal cord. Their axons leave the brain stem by way of cranial nerves III, VII, IX, X, and XI and leave the cord by some pelvic nerves. They extend some distance before terminating in the parasympathetic ganglia located in the head and in the thoracic and abdominal cavities close to visceral effectors. The dendrites and cell bodies of parasympathetic postganglionic neurons lie in these outlying parasympathetic ganglia and their short axons extend into the nearby structures. Each parasympathetic preganglionic neuron synapses, therefore, only with postganglionic neurons to a single effector. For this reason, parasympathetic stimulation frequently involves response by only one organ. This is not true of sympathetic responses. As we have noted, sympathetic stimulation usually results in responses by numerous organs.

Function

The parasympathetic system dominates control of many visceral effectors under normal, everyday conditions. Impulses over parasympathetic fibers, for example, tend to slow the heartbeat, increase peristalsis, and increase secretion of digestive juices and insulin.

Autonomic conduction paths

Conduction paths to visceral and somatic effectors from the central nervous system (spinal cord or brain stem) differ somewhat. Autonomic paths to visceral effectors, as the right side of Fig. 7-1 shows, consist of two-neuron relays. Impulses travel over preganglionic neurons from the cord or brain stem to autonomic ganglia. Here they are relayed across synapses to postganglionic neurons, which then conduct them from the ganglia to visceral effectors. Compare this with the somatic conduction path

illustrated on the left side of Fig. 7-1. Somatic motoneurons, like the ones shown here, conduct all the way from cord or brain stem to somatic effectors with no intervening synapses.

Autonomic neurotransmitters

Turn your attention now to Fig. 7-2. It reveals information about autonomic neurotransmitters, the chemical compounds released from the axon terminals of autonomic neurons. Observe that three of the axons shown in Fig. 7-2, namely, the sympathetic preganglionic axon, the parasympathetic preganglionic axon, and the parasympathetic postganglionic axon, release acetylcholine. These axons are therefore classified as *cholinergic fibers*. Only one type of autonomic axons releases the neurotransmitter norepinephrine. These are the axons of sympathetic postganglionic neurons and they are classified as *adrenergic fibers*. See p. 88 for other neurons whose axons release acetylcholine and norepinephrine.

Autonomic nervous system as a whole

The function of the autonomic nervous system is to regulate the body's automatic, nonwilled functions in ways that tend to maintain or quickly restore homeostasis. Many internal organs are doubly innervated by the autonomic nervous system. In other words, they receive fibers from parasympathetic and sympathetic divisions. Both parasympathetic and sympathetic impulses continually bombard them and, as Table 7-1 indicates, influence their functioning in opposite or antagonistic ways. For example, the heart continually receives sympa-

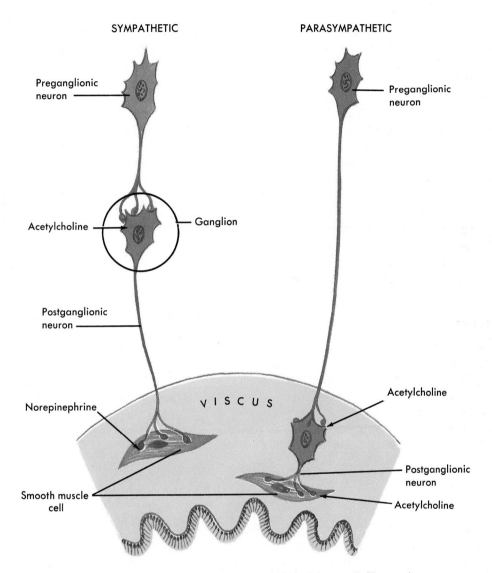

Fig. 7-2 Cholinergic fibers release acetycholine (ACh). Adrenergic fibers release norepinephrine (NE). Note the three kinds of cholinergic fibers shown in the diagram: axons of both sympathetic and parasympathetic preganglionic neurons and of which postganglionic neurons? Which postganglionic neurons would you classify as adrenergic? Almost immediately after its release, ACh is inactivated by the enzyme acetylcholinesterase (AChE). The enzyme catechol-O-methyl transferase (COMT) inactivates NE.

thetic impulses that tend to make it beat faster and parasympathetic impulses that tend to slow it down. The ratio between these two antagonistic forces determines the actual heart rate.

The name "autonomic nervous system" is something of a misnomer. It seems to imply that this part of the nervous system is independent from other parts. This is not true. Dendrites and cell bodies of preganglionic neurons are located, as we have observed, in the spinal cord and the brain stem. They are continually influenced directly or indirectly by impulses from neurons located above them, notably by some in the hypothalamus and in the parts of the cerebral cortex called the "emotional brain." Through conduction paths from these areas, emotions can produce widespread changes in the automatic functions of our bodies, in cardiac and smooth muscle contractions, and in secretion by glands. Anger and fear, for example, lead to increased sympathetic activity and the fight-or-flight syndrome. And according to some physiologists, the altered state of consciousness known as meditation or yoga leads to decreased sympathetic activity and a group of changes opposite to those of the fight-or-flight syndrome.

Outline summary

Definitions

A Autonomic nervous system—motoneurons that conduct impulses from the central nervous system to cardiac muscle, smooth muscle, and glandular epithelial tissue; nervous system structures that regulate the body's automatic or involuntary functions

B Autonomic neurons—motoneurons that conduct from the central nervous system to cardiac muscle, smooth muscle, and glandular epithelial tissue; preganglionic autonomic neurons that conduct from spinal cord or brain stem to an autonomic ganglion; post-

ganglionic neurons that conduct from autonomic ganglia to cardiac muscle, smooth muscle, and glandular epithelial tissue

C Autonomic or visceral effectors—tissues to which autonomic neurons conduct impulses, that is, cardiac and smooth muscle and glandular epithelial tissue

Names of divisions

A Sympathetic or thoracolumbar system

B Parasympathetic or craniosacral system

Sympathetic nervous system

A Structure
  1 Dendrites and cell bodies of sympathetic preganglionic neurons located in gray matter of thoracic and upper lumbar segments of cord
  2 Axons leave cord in anterior roots of spinal nerves, extend to sympathetic or collateral ganglia, and synapse with several postganglionic neurons whose axons extend out in spinal nerves or autonomic nerves to terminate in visceral effectors
  3 Chain of sympathetic ganglia in front of and at each side of spinal column

B Functions
  1 Serves as the emergency or stress system, controlling visceral effectors during strenuous exercise and strong emotions (anger, fear, hate, anxiety)
  2 Group of changes induced by sympathetic control is called the fight-or-flight syndrome

Parasympathetic nervous system

A Structure
  1 Parasympathetic preganglionic neurons have their dendrites and cell bodies in gray matter of the brain stem and of the sacral segments of cord
  2 Their axons leave brain stem by way of cranial nerves III, VII, IX, X, and XI and leave the cord by some pelvic nerves
  3 They terminate in parasympathetic ganglia located in head and thoracic and abdominal cavities close to visceral effectors
  4 Each parasympathetic preganglionic neuron synapses with postganglionic neurons to only one effector

B Function—dominates control of many visceral effectors under normal, everyday conditions

Autonomic conduction paths

A Consist of two-neuron relays, that is, preganglionic neurons from central nervous system to autonomic ganglia, synapses, postganglionic neurons from ganglia to visceral effectors

B In contrast, somatic motoneurons conduct all the way from central nervous system to somatic effectors with no intervening synapses.

Autonomic neurotransmitters

A Cholingeric fibers—preganglionic axons of both parasympathetic and sympathetic systems and parasympathetic postganglionic axons release acetylcholine

B Adrenergic fibers—axons of sympathetic postganglionic neurons release norepinephrine (noradrenalin)

C Acetylcholine and norepinephrine are not exclusively autonomic neurotransmitters; for example, axons of somatic motoneurons release acetylcholine and axons of various neurons located in central nervous system are now known to release norepinephrine

Autonomic nervous system as a whole

A Regulates the body's automatic functions in ways that tend to maintain or quickly restore homeostasis

B Many visceral effectors are doubly innervated, that is, receive fibers from both parasympathetic and sympathetic divisions and are influenced in opposite ways by the two divisions

C Autonomic nervous system is not autonomous or independent from other parts of the nervous system; conduction paths from hypothalamus and other parts of the brain influence conduction by both parasympathetic and sympathetic divisions

Review questions

1 Contrast the meanings of the terms "visceral effectors" and "somatic effectors."
2 Differentiate between a preganglionic neuron and a postganglionic neuron.
3 Compare sympathetic and parasympathetic systems as to locations of their preganglionic dendrites, cell bodies, and axons and their postganglionic dendrites, cell bodies, and axons.
4 All preganglionic neurons release the same neurotransmitter. Name it. Only one kind of autonomic axon releases norepinephrine. Which kind?
5 Compare parasympathetic and sympathetic functions.
6 Explain why the name "autonomic nervous system" is misleading.
7 Explain the meaning of the term "fight-or-flight syndrome."
8 Which division of the autonomic nervous system functions as an emergency or stress system?
9 What general function does the autonomic nervous system perform?

chapter 8

The endocrine system

Have you ever seen a giant or a dwarf? Have you ever known anyone who had "sugar diabetes" or a goiter? If so, you have had visible proof of the importance of the endocrine system for normal development and health.

The endocrine system performs the same general functions as the nervous system, namely, communication and control. The nervous system provides rapid, brief control by fast-traveling nerve impulses. The endocrine system provides slower but longer-lasting control by hormones (chemicals) secreted into and circulated by the blood.

The organs of the endocrine system are located in widely separated parts of the body—in the cranial cavity, in the neck, in the thoracic cavity, in the abdominal cavity, in the pelvic cavity, and outside the body cavities. Note the names and locations of the endocrine glands shown in Fig. 8-1 and Table 8-1.

All the organs of the endocrine system are glands, but all glands are not organs of the endocrine system. Those glands that discharge their secretions into ducts are not endocrine glands. They are exocrine or duct glands. *Endocrine glands* are ductless glands. They secrete chemicals known as *hormones* into the blood. Cells that are acted on and respond in some way to a particular hormone are referred to descriptively as *target organ cells*. The list of endocrine glands and their target organs continues to grow. Only the names and lo-

112

Table 8-1 Names and locations of endocrine glands	
Name	**Location**
Pituitary gland Anterior lobe	Cranial cavity
Posterior lobe	
Thyroid gland	Neck
Parathyroid glands	Neck
Adrenal glands Adrenal cortex	Abdominal cavity (retroperitoneal)
Adrenal medulla	
Islands of Langerhans	Abdominal cavity (pancreas)
Ovaries Ovarian follicle	Pelvic cavity
Corpus luteum	
Testes (interstitial cells)	Scrotum
Thymus	Mediastinum
Placenta	Pregnant uterus

cations of the main endocrine glands are given in Table 8-1 and Fig. 8-1.

In this chapter you will read about the functions of the main endocrine glands and discover why their importance is almost impossible to exaggerate. Hormones are the main regulators of metabolism, growth and development, reproduction, and many other body activities. They play roles of the utmost importance in maintaining homeostasis—fluid and electrolyte balance, acid-base balance, and energy balance, for example. Hormones make the difference between normalcy and all sorts of abnormalities such as dwarfism, giantism, and sterility. They are important not only for the healthy survival of each one of us, but also for the survival of the human species itself.

Prostaglandins

Prostaglandins (PG's) or tissue hormones are important and extremely powerful substances that have been found in a wide variety of tissues in the body. They play an important role in communication and control of many body functions but do not meet the definition of a typical hormone. The term "tissue hormone" is appropriate because in many instances a prostaglandin is produced in a tissue and diffuses only a short distance to act on cells within that tissue. Typical hormones influence and control activities of widely separated organs; typical prostaglandins influence activities of neighboring cells.

The prostaglandins in the body can be divided into several groups. Three classes of prostaglandins, prostaglandin A (PGA),

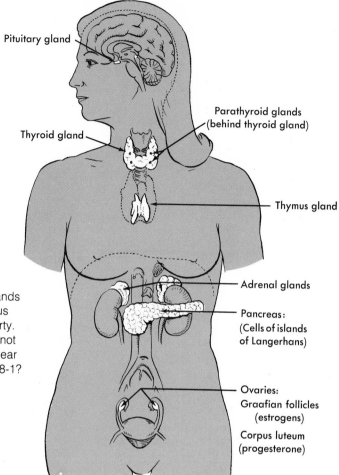

Pituitary gland

Parathyroid glands
(behind thyroid gland)

Thyroid gland

Thymus gland

Adrenal glands

Pancreas:
(Cells of islands
of Langerhans)

Ovaries:
Graafian follicles
(estrogens)

Corpus luteum
(progesterone)

Fig. 8-1 Location of the endocrine glands in the female. Dotted line around thymus gland indicates maximum size at puberty. Which glands appear in this figure but not in Table 8-1? What gland does not appear in this figure but does appear in Table 8-1?

prostaglandin E (PGE), and prostaglandin F (PGF) are among the most common. Although we do not know the exact mechanism by which prostaglandins act on cells, they have profound effects on many body functions. They influence respiration, blood pressure, gastrointestinal secretions, and the reproductive system. Eventually, prostaglandins may play an important role in the treatment of such diverse diseases as high blood pressure, asthma, and ulcers.

Pituitary gland

The pituitary gland (hypophysis cerebri) is truly a small but mighty structure. Although no larger than a pea, it is really two endocrine glands. One is called the anterior pituitary gland (or adenohypophysis) and the other is called the posterior pituitary gland (or neurohypophysis). Differences between the two glands are suggested by their names—*adeno* means "gland" and *neuro* means "nervous." The adenohypophysis has

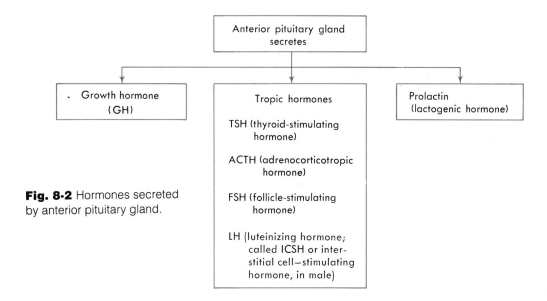

Fig. 8-2 Hormones secreted by anterior pituitary gland.

the structure of an endocrine gland, whereas the neurohypophysis has the structure of nervous tissue. Hormones secreted by the adenohypophysis serve very different functions from those released from the neurohypophysis.

The protected location of this dual gland suggests its importance. The pituitary gland lies buried deep in the cranial cavity, in the small depression of the sphenoid bone that is shaped like a saddle and is called the *sella turcica* (Turkish saddle.) A stemlike structure, the pituitary stalk, attaches the gland to the undersurface of the brain. More specifically, the stalk attaches the pituitary body to the hypothalamus.

Anterior pituitary gland hormones

The anterior pituitary gland secretes six major hormones. To learn their abbreviated and full names, see Fig. 8-2. Each of the four hormones listed as a *tropic hormone* stimulates another endocrine gland to grow and secrete its hormones. Because the anterior pituitary gland exerts this control over the structure and function of the thyroid gland, the adrenal cortex, the ovarian follicles, and the corpus luteum, it is often referred to as the *master gland.*

Thyroid-stimulating hormone (TSH) acts on the thyroid gland. As its name suggests, it stimulates the thyroid gland to increase its secretion of thyroid hormone.

Adrenocorticotropic hormone (ACTH) acts on the adrenal cortex. It stimulates the adrenal cortex to increase in size and to secrete larger amounts of its hormones, especially of cortisol (hydrocortisone).

Follicle-stimulating hormone (FSH) stimulates primary ovarian follicles in an ovary to start growing and to continue developing to maturity, that is, to the point of ovulation. FSH also stimulates follicle cells to secrete estrogens. In the male, FSH stimulates the seminiferous tubules to grow and form sperm.

Luteinizing hormone (LH) acts with FSH to perform four functions. It stimulates a follicle and ovum to complete their growth to maturity. It stimulates follicle cells to secrete estrogens. It causes ovulation (rupturing of the mature follicle with expulsion of its ripe ovum). Because of this function, LH is sometimes called the ovulating hormone. Finally, LH stimulates the formation of a golden body, the corpus luteum, in the ruptured follicle—the process is called luteinization. This function, obviously, is the one that earned LH its title of luteinizing hormone. The male pituitary gland also secretes LH, but it is called *interstitial cell–stimulating hormone* (ICSH) because it stimulates interstitial cells in the testes to develop and secrete testosterone, the male sex hormone (Fig. 8-4).

Melanocyte-stimulating hormone (MSH) causes a rapid increase in the synthesis and dispersion of melanin (pigment) granules in specialized skin cells (Fig. 8-4).

The other two important hormones secreted by the anterior pituitary gland are growth hormone and prolactin (lactogenic hormone). Growth hormone (GH) tends to speed up the movement of digested proteins (amino acids) out of the blood and into the cells, and this accelerates the cells' anabolism of amino acids to form tissue proteins; hence this action promotes normal growth. Growth hormone also affects fat and carbohydrate metabolism: it accelerates fat catabolism but slows glucose catabolism. This means that less glucose leaves the blood to enter cells and therefore the amount of glucose in the blood tends to increase. Thus growth hormone and insulin have opposite effects on blood glucose. Insulin tends to decrease blood glucose, and growth hormone tends to increase it. Too much insulin in the blood produces *hypoglycemia* (lower than normal blood glucose concentration). Too much growth hormone produces *hyperglycemia* (higher than normal blood glucose concentration). This type of hyperglycemia is appropriately called pituitary diabetes,

and growth hormone is referred to as a diabetogenic (diabetes-causing) or hyperglycemic hormone.

During pregnancy, prolactin stimulates the breast development necessary for eventual lactation (milk secretion). Also, soon after delivery of a baby, prolactin stimulates the breasts to start secreting milk, a function suggested by prolactin's other name, lactogenic hormone.

For a brief summary of anterior pituitary hormone functions, see Figs. 8-3 and 8-4.

Posterior pituitary gland hormones

The posterior pituitary gland releases two hormones—antidiuretic hormone (ADH) and oxytocin. ADH accelerates the reabsorption of water from urine in kidney tubules back into the blood. With more water moving out of the tubules into the blood, less water remains in the tubules, and therefore less urine leaves the body. The reason the name "antidiuretic hormone" is an appropriate one is that *anti-* means "against" and *diuretic* means "increasing the volume of urine excreted." Therefore antidiuretic means "acting against an increase in urine volume"; in other words, the antidiuretic hormone (ADH) acts to decrease urine volume.

The posterior pituitary hormone oxytocin is secreted by a woman's body before and after she has a baby. Oxytocin stimulates contraction of the smooth muscle of the pregnant uterus and so is believed to initiate and maintain labor. This is why physicians sometimes prescribe oxytocin injections to induce or increase labor. Oxytocin also performs a function that is important to a newborn baby. It causes the glandular cells of the breast to release milk into ducts from which a baby can obtain it by sucking. Fig. 8-5 summarizes posterior pituitary functions.

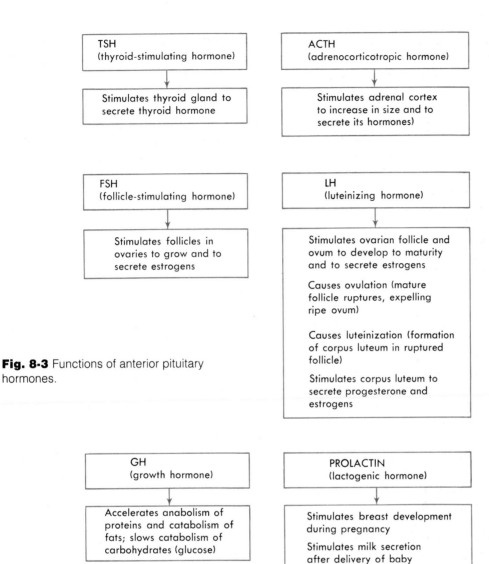

Fig. 8-3 Functions of anterior pituitary hormones.

Thyroid gland

Earlier in this chapter we mentioned that some endocrine glands are not located in a body cavity. The thyroid is one of these. It lies in the neck just below the larynx (Fig. 8-6).

The thyroid gland secretes two hormones—thyroid hormone and calcitonin.

The thyroid hormone influences every one of the trillions of cells in our bodies. It makes them speed up their release of energy from foods. In other words, thyroid hormone stimulates cellular catabolism. This has far-reaching effects. Because all body functions depend on a normal supply of energy, they all depend on normal thyroid secretion. Even normal mental development and normal physical growth and develop-

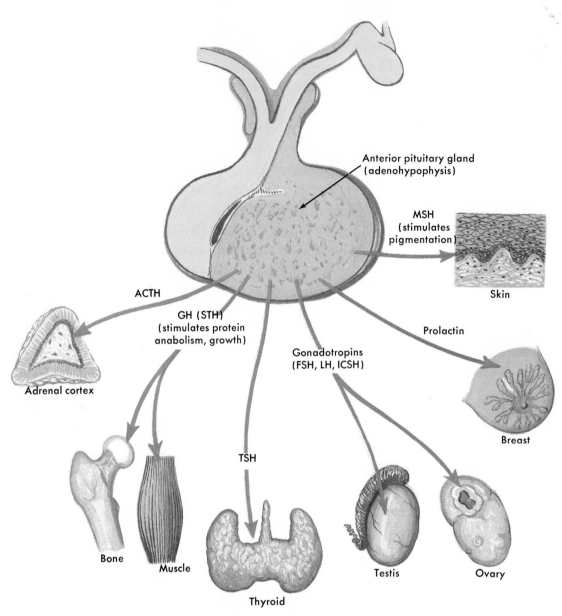

Fig. 8-4 Anterior pituitary hormones and their target organs: adrenocorticotropic hormone (ACTH); thyroid-stimulating hormone (TSH); follicle-stimulating hormone (FSH); luteinizing hormone (LH); male analog of LH (ICSH); melanocyte-stimulating hormone (MSH).

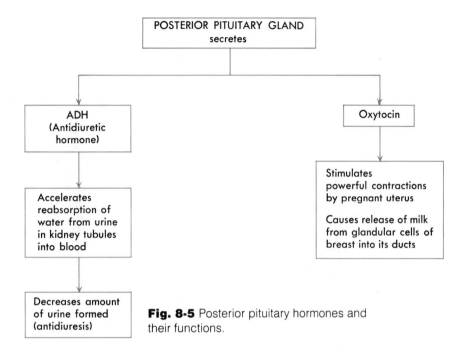

Fig. 8-5 Posterior pituitary hormones and their functions.

ment depend on normal thyroid functioning.

If the thyroid secretes too little hormone, catabolism slows down. As a result, cells have too little energy and cannot do their work properly. If the thyroid secretes too much hormone, catabolism speeds up too much. This produces very noticeable effects. The hyperthyroid individual appears nervous, jumpy, and excessively active—hyperactive, we say.

Calcitonin tends to decrease the concentration of calcium in the blood, probably by first acting on bone to inhibit its breakdown. With less bone being destroyed, less calcium moves out of bone into blood and, as a result, the concentration of calcium in blood decreases. An increase in calcitonin secretion quickly follows any increase in blood calcium concentration even if it is a slight one. This causes blood calcium concentration to decrease back to its normal lower level. Calcitonin thus functions to help maintain homeostasis of blood calcium. It prevents a harmful excess of calcium in the blood (hypercalcemia) from developing.

Parathyroid glands

The parathyroid glands are small glands. There are usually four of them, and they are found on the back of the thyroid gland (Fig. 8-6). The parathyroid glands secrete parathyroid hormone (PTH).

Parathyroid hormone tends to increase the concentration of calcium in the blood—the opposite effect of the thyroid gland's calcitonin. Whereas calcitonin acts to decrease the amount of calcium being resorbed from bone, parathyroid hormone acts to increase it. Parathyroid hormone stimulates bone-destroying cells to increase their breakdown of bone's hard matrix, a process that frees the calcium stored in the

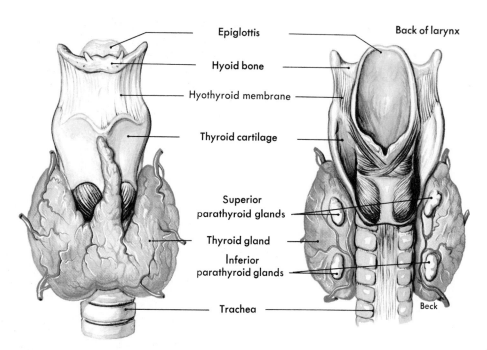

Fig. 8-6 Thyroid and parathyroid glands. Note their relations to each other and to the larynx (voice box) and trachea.

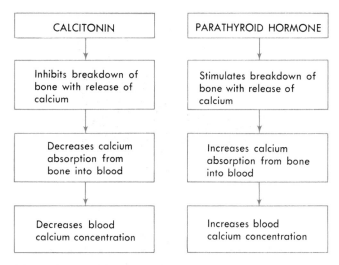

Fig. 8-7 Effects of blood calcium concentration of calcitonin and parathyroid hormones.

matrix. The released calcium then moves out of bone into blood, and this in turn increases blood's calcium concentration. For a summary of the antagonistic effects of calcitonin and parathyroid hormone, see Fig. 8-7. This is a matter of life-and-death importance because our cells are extremely sensitive to changing amounts of blood calcium. They cannot function normally either with too much or with too little calcium. For example, with too much blood calcium, brain cells and heart cells soon do not function normally; a person becomes mentally disturbed and his heart may stop. However, with too little blood calcium, nerve cells become overactive—sometimes to such an extreme degree that they bombard muscles with so many impulses that

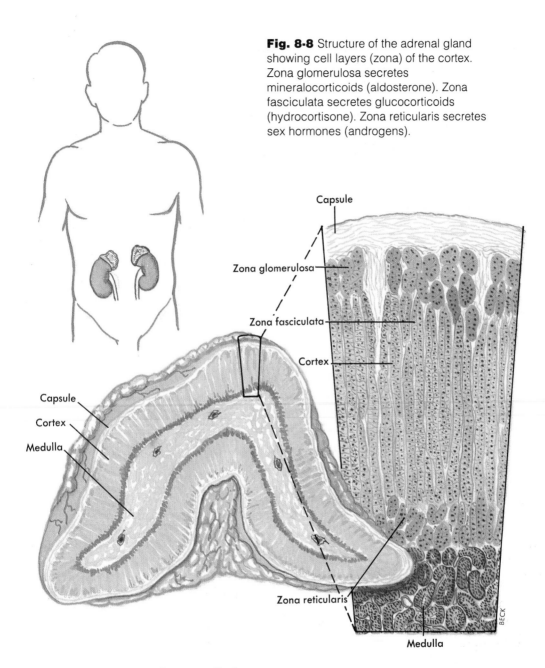

Fig. 8-8 Structure of the adrenal gland showing cell layers (zona) of the cortex. Zona glomerulosa secretes mineralocorticoids (aldosterone). Zona fasciculata secretes glucocorticoids (hydrocortisone). Zona reticularis secretes sex hormones (androgens).

Capsule

Zona glomerulosa

Zona fasciculata

Cortex

Capsule

Cortex

Medulla

Zona reticularis

Medulla

the muscles go into spasms. This is called *tetany*.

Based on what you have just read about calcitonin and parathyroid hormone, which condition might lead to too little blood calcium and tetany—a deficiency or an excess of parathyroid hormone?

Adrenal glands

As you can see in Figs. 8-1 and 8-8, an adrenal gland curves over the top of each kidney. Although from the surface an adrenal gland appears to be only one organ, it actually is two separate endocrine glands, namely, the adrenal cortex and the adrenal medulla. Does this two-glands-in-one structure remind you of another endocrine organ? (See p. 114). The adrenal cortex is the outer part of an adrenal gland and the medulla is its inner part. Adrenal cortex hormones have different names and quite different actions from adrenal medulla hormones.

Adrenal cortex

Three different zones or layers of cells make up the adrenal cortex (Fig. 8-8). Starting with the zone or layer directly under the outer capsule of the gland, their names are *zona glomerulosa, zona fasciculata,* and *zona reticularis.* Hormones secreted by the three cell layers or zona of the adrenal cortex are called *corticoids.* The outer zone of adrenal cortex cells (zona glomerulosa) secretes hormones called mineralocorticoids (or MC's, for short). The main mineralocorticoid is aldosterone. The middle zone or layer (zona fasciculata) secretes glucocorticoids (or GC's). Cortisol or hydrocortisone is the chief glucocorticoid. The innermost or deepest zone of the cortex (zona reticularis) secretes small amounts of sex hormones. Sex hormones secreted by the adrenal cortex resemble testosterone. We shall now discuss briefly the functions of these three kinds of adrenal cortical hormones.

One of the important functions of glucocorticoids is to help maintain normal blood glucose concentration. Glucocorticoids tend to increase gluconeogenesis, a process that converts amino acids or fatty acids to glucose and that is carried on mainly by liver cells. Glucocorticoids act in several ways to increase gluconeogenesis. They promote the breakdown to amino acids of tissue proteins, especially those present in muscle cells. The amino acids thus formed move out of the tissue cells into blood and circulate to the liver. Liver cells then change

them to glucose by the process of gluconeogenesis. The newly formed glucose leaves the liver cells and enters the blood. This of course increases blood glucose concentration.

In addition to performing these functions that are necessary for maintaining normal blood glucose concentration, glucocorticoids also play an essential part in maintaining normal blood pressure. They act in a complicated way to make it possible for two other hormones (epinephrine and norepinephrine secreted by the medulla of the adrenal gland) to partially constrict blood vessels, a condition necessary for maintaining normal blood pressure. Also, glucocorticoids act with these hormones from the adrenal medulla to produce an anti-inflammatory effect. They bring about a normal recovery from inflammations produced by many kinds of agents. The use of hydrocortisone to relieve rheumatoid arthritis, for example, is based on the anti-inflammatory effect of glucocorticoids.

Another effect produced by glucocorticoids is called their anti-immunity, antiallergy effect. Glucocorticoids bring about a decrease in the number of certain cells (lymphocytes and plasma cells) that produce antibodies, substances that make us immune to some factors and allergic to others.

When extreme stimuli act on the body, they produce an internal state or condition known as *stress.* Surgery, hermorrhage, infections, severe burns, and intense emotions are examples of extreme stimuli that bring on stress. The normal adrenal cortex responds to the condition of stress by quickly increasing its secretion of glucocorticoids. This fact is well established. What is still not known, however, is whether the increased amount of glucocorticoids helps the body cope successfully with stress. Increased glucocorticoid secretion is only one of many responses that the body makes to stress, but it is one of the first stress responses and it brings about many of the other stress responses. Examine Fig. 8-9 to discover what stress responses are produced by a high concentration of glucocorticoids in the blood.

Mineralocorticoids perform different functions from glucocorticoids. As their name suggests, they help control the amount of certain mineral salts in the blood. Aldosterone is the name of the chief mineralocorticoid. Remember its main functions—to increase the amount of sodium and decrease the amount of potassium in the blood—for these changes lead to other profound changes. Aldosterone increases blood sodium and decreases blood potassium by influencing the kidney tubules. It causes them to speed up their reabsorption of sodium back into the blood so that less of it will be lost in the urine. At the same time, aldosterone causes the tubules to increase their secretion of potassium so that more of this

mineral will be lost in the urine. Aldosterone also tends to speed up kidney reabsorption of water.

One kind of sex hormone secreted by the adrenal cortex acts like the male sex hormone testosterone. Compounds of this type therefore are called *androgens* (from the Greek combining form *andro-*, meaning "man"). Androgens are male sex hormones; they produce masculinizing effects. Strangely enough, even a woman's adrenal glands secrete some androgens. And the adrenal cortex of both sexes secretes female sex hormones, but usually in such small amounts that they produce no noticeable effects.

Adrenal medulla

The adrenal medulla, or inner portion of the adrenal glands, secretes *epinephrine* and *norepinephrine*. (Biochemists refer to these compounds as *catecholamines*.)

Our bodies have many ways to defend themselves against enemies that threaten

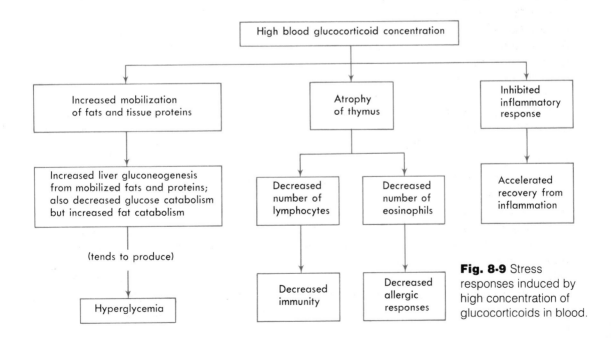

Fig. 8-9 Stress responses induced by high concentration of glucocorticoids in blood.

their well-being. A physiologist might state this same truth by saying that the body resists stress by making many stress responses. We have just discussed increased glucocorticoid secretion. An even faster-acting stress response is increased secretion by the adrenal medulla. (It occurs very rapidly because nerve impulses conducted by sympathetic nerve fibers stimulate the adrenal medulla.) When stimulated, it literally squirts epinephrine and norepinephrine into the blood. Like the glucocorticoids, these hormones may help the body resist stress. Unlike the glucocorticoids, they are not essential for maintaining life. Glucocorticoids, the hormones from the adrenal cortex, on the other hand, may help the body resist stress and are essential for life.

Suppose you suddenly faced some threatening situation. Imagine that you found a lump in your breast, or that your doctor told you that you had to have a dangerous operation, or that a gunman threatened to kill you. Almost instantaneously, the medullas of your two adrenal glands would be galvanized into feverish activity. They would quickly secrete large amounts of epinephrine (adrenaline) into your blood. Many of your body functions would seem to become supercharged. Your heart would beat faster; your blood pressure would rise; more blood would be pumped to your skeletal muscles; your blood would contain more glucose for more energy, and so on. In short, you would be geared for strenuous activity, for "fight or flight." Epinephrine prolongs and intensifies changes in body function brought about by one divison of the nervous system. (Do you recall which one? Check your answer on p. 108).

Islands of Langerhans

All of the endocrine glands discussed so far are big enough to be seen without a magnifying glass or microscope. The islands of Langerhans, in contrast, are too tiny to be seen without a microscope. These glands are merely little clumps of cells scattered like islands in a sea among the pancreatic cells that secrete the pancreatic digestive juice. Paul Langerhans discovered these cell islands in the pancreas about 100 years ago—hence their name islands of Langerhans.

Two kinds of cells in the islands of Langerhans are those called alpha and beta cells. Alpha cells secrete a hormone called *glucagon*, whereas beta cells secrete one of the most famous of all hormones, *insulin*.

Glucagon accelerates a process called *liver glycogenolysis*. Glycogenolysis is a chemical process by which the glucose stored in the liver cells in the form of glycogen is converted to glucose. This glucose then leaves the liver cells and enters the blood. Glucagon, therefore, tends to increase blood glucose concentration.

Insulin and glucagon serve as antagonists. In other words, insulin tends to decrease blood glucose concentration; glucagon tends to increase it. Insulin is the only hormone that can decrease blood glucose concentration. Several other hormones, however, tend to increase its concentration. We have already named three of these: glucocorticoids, growth hormone, and glucagon. Insulin decreases blood glucose by accelerating its movement out of the blood, through cell membranes, and into cells. As glucose enters the cells at a faster rate, the cells increase their metabolism of glucose. Briefly then, insulin functions to decrease blood glucose and to increase glucose metabolism.

If the islands of Langerhans secrete a normal amount of insulin, a normal amount of glucose enters the cells, and a normal amount of glucose stays behind in the blood. ("Normal" blood glucose is about 80 to 120 milligrams of glucose in every 100 milliliters of blood.) If the islands of Langerhans secrete too much insulin, as they sometimes do when a person has a tumor of the pancreas, then more glucose than usual leaves the blood to enter the cells and blood glucose decreases. If the islands of Langerhans secrete too little insulin, as they do in *diabetes mellitus,* less glucose leaves the blood to enter the cells so that blood glucose increases—sometimes to even three or more times the normal amount.

Female sex glands

A woman's sex glands are her two ovaries. Each ovary contains two different kinds of glandular structures, namely, the ovarian follicles and the corpus luteum. *Ovarian follicles* secrete estrogens, the "feminizing hormone." The *corpus luteum* secretes chiefly progesterone but also some estrogens. We shall save our discussion of the structure of these endocrine glands and the functions of their hormones for Chapter 16.

Male sex glands

Some of the cells of the testes secrete semen (the male reproductive fluid) into ducts. Other cells, referred to as interstitial cells, secrete the male hormone testosterone into the blood. The interstitial cells of the testes therefore are the male endocrine glands. Testosterone is the "masculinizing hormone." Chapter 15 contains more information about the structure of the testes and the functions of testosterone.

Thymus

The thymus is located in the mediastinum (Fig. 8-1), and in infants it may extend up into the neck as far as the lower edge of the thyroid gland. Like the adrenal gland, the thymus has a cortex and medulla. Both portions are composed largely of lymphocytes. There is no longer any doubt that the thymus functions as an endocrine gland. As an important part of the body's immune system, the endocrine function of the thymus is not only important but essential. This small structure (it weighs at most a little over an ounce) plays a critical part in the body's defenses against infections—in its vital immunity mechanism.

The hormone *thymosin* has been isolated from thymus tissue and is considered responsible for its endocrine activity. Thymosin is actually a group of several hormone-like substances that together play an important role in the development and functioning of the body's immune system.

Suppression of the immune system sometimes occurs in certain disease states and in patients who are undergoing massive chemotherapy or radiotherapy for the treatment of cancer. Such individuals are said to be "immunosuppressed" and are extremely susceptible to infections. Thymosin may prove useful as an activator of the immune system in such patients.

Placenta

The placenta functions as a temporary endocrine gland. During pregnancy, it produces *chorionic gonadotropins,* so-called because they are secreted by cells of the *chorion,* the outermost membrane that surrounds the baby during development in the uterus. In addition to chorionic gonadotropins, the placenta also produces estrogen and progesterone. During pregnancy, the kidneys excrete large amounts of chorionic gonadotropins in the urine. This fact, discovered more than a half century ago by Aschheim and Zondek, led to the development of the now-familiar pregnancy tests.

Endocrine diseases

Diseases of the endocrine glands are numerous, varied, and sometimes spectacular. Tumors or other abnormalities frequently cause the glands to secrete either too much (hypersecretion) or too little (hyposecretion) of their hormones. We shall identify a few of the more common conditions characterized by hypersecretion or hyposecretion of hormones.

Excessive secretion of growth hormone during the early years of life produces a condition called *giantism*. The name suggests the obvious characteristics of this condition. The child grows to giant size. If the anterior pituitary gland secretes too much growth hormone after the growth years, then the disease called *acromegaly* develops. The individual's bones—particularly of the face, hands, and feet—enlarge gradually but strikingly, often changing the person's appearance enough to make him almost unrecognizable. Excess growth hormone tends to produce excess blood glucose (hyperglycemia or "pituitary diabetes").

Almost everyone has known someone with a hyperactive thyroid gland, or, as it is commonly called, a *goiter*. Among the chief symptoms of a hyperactive thyroid are an increased metabolic rate, a very rapid heartbeat, and nervous excitability.

Hyposecretion of thyroid hormone during the early years of childhood leads to a disease called *cretinism*. Malformed dwarfs are victims of cretinism. Their metabolic rate is low, and frequently they show signs of retarded mental, physical, and sexual development. If the thyroid functions normally during the growth years and then starts to secrete too little of its hormone, the disease *myxedema* results. Besides a low metabolic rate, persons with myxedema usually show signs of lessened mental and physical vigor. They generally gain weight and often complain that their hair is falling out.

Cushing's syndrome is the medical name for the hypersecretion of glucocorticoids. For some reason many more women than men develop Cushing's syndrome. Its most noticeable features are the so-called moon face and buffalo hump that develop because of the redistribution of body fat.

A deficiency of insulin secretion leads to an excess of blood glucose, the identifying characteristic of *diabetes mellitus*, probably the best known of all endocrine disorders. A higher than normal blood glucose in turn leads to many other symptoms and, if high enough for long, can cause death.

Deficient secretion of the antidiuretic hormone by the posterior pituitary gland causes an excess secretion of urine, that is, *diuresis*, the identifying characteristics of the disease called *diabetes insipidus*.

Outline summary
General functions

A Endocrine glands secrete chemicals (hormones) into the blood
B Hormones perform general functions of communication and control—but a slower, longer-lasting type of control than that provided by nerve impulses
C Cells acted on by hormones are called target organ cells

Prostaglandins

A Prostaglandins (PG's) are powerful substances found in a wide variety of body tissues
B PG's are often produced in a tissue and diffuse only a short distance to act on cells in that tissue
C Several classes of prostaglandins, including prostaglandin A (PGA), prostaglandin E (PGE), and prostaglandin F (PGF)
D PG's influence many body functions including respiration, blood pressure, gastrointestinal secretions, and reproduction

Pituitary gland

A Anterior pituitary gland (adenohypophysis)
  1 Names of major hormones
   a Thyroid-stimulating hormone (TSH)
   b Adrenocorticotropic hormone (ACTH)
   c Follicle-stimulating hormone (FSH)
   d Luteinizing hormone (LH)
   e Interstitial cell–stimulating hormone (ICSH)

f Melanocyte-stimulating hormone (MSH)
g Growth hormone (GH)
h Prolactin (lactogenic hormone)
2 Functions of major hormones
a TSH—stimulates growth of thyroid gland; also stimulates it to secrete thyroid hormone
b ACTH—stimulates growth of adrenal cortex and stimulates it to secrete glucocorticoids (mainly cortisol)
c FSH—initiates growth of ovarian follicles each month in ovary and stimulates one or more follicles to develop to stage of maturity and ovulation; FSH also stimulates estrogen secretion by developing follicles
d LH—acts with FSH to stimulate estrogen secretion and follicle growth to maturity; LH causes ovulation and is "the ovulating hormone"; LH causes luteinization of the ruptured follicle and stimulates progesterone secretion by corpus luteum
e ICSH—male pituitary secretes LH but it is called interstitial cell–stimulating hormone (ICSH); causes interstitial cells in testes to secrete testosterone
f MSH—causes rapid increase in synthesis and spread of melanin (pigment) in the skin
g GH—stimulates growth by accelerating protein anabolism; also, tends to accelerate fat catabolism and slow glucose catabolism; by slowing glucose catabolism, GH tends to increase blood glucose to higher than normal level (hyperglycemia)
h Prolactin (lactogenic hormone)—stimulates breast development during pregnancy and secretion of milk after delivery of baby
B Posterior pituitary gland (neurohypophysis)
1 Names of hormones
a Antidiuretic hormone (ADH)
b Oxytocin
2 Functions of hormones
a ADH—accelerates water reabsorption from urine in kidney tubules into blood, thereby decreasing urine secretion (antidiuresis)
b Oxytocin—stimulates pregnant uterus to contract; may initiate labor; causes glandular cells of breast to release milk into ducts

Thyroid gland

A Names of hormones
1 Thyroid hormone
2 Calcitonin
B Functions of hormones
1 Thyroid hormone—accelerates catabolism (tends to increase basal metabolic rate)
2 Calcitonin—tends to decrease blood calcium concentration, perhaps by inhibiting breakdown of bone with release of calcium into blood

Parathyroid glands

A Name of hormone—parathyroid hormone
B Function of hormone—hormone tends to increase blood calcium concentration by increasing breakdown of bone with release of calcium into blood

Adrenal glands

A Adrenal cortex
1 Names of hormones (corticoids)
a Glucocorticoids (GC's)—chiefly cortisol (hydrocortisone)
b Mineralocorticoids (MC's)—chiefly aldosterone
c Sex hormones—small amounts of male hormones (androgens) and female hormone (estrogens) secreted by adrenal cortex of both sexes
2 Cell layers (zonae) (Fig. 8-8)
a Zona glomerulosa—outermost layer, secretes mineralocorticoids
b Zona fasciculata—middle layer, secretes glucocorticoids
c Zona reticularis—deepest or innermost layer, secretes sex hormones
3 Functions of glucocorticoids
a Help maintain normal blood glucose concentration by increasing gluconeogenesis—the formation of "new" glucose from amino acids produced by the breakdown of proteins, mainly those in muscle tissue cells; also the conversion to glucose of fatty acids produced by the breakdown of fats stored in adipose tissue cells
b Play an essential part in maintaining normal blood pressure—glucocorticoids make it possible for epinephrine and norepinephrine to maintain a normal degree of vasoconstriction, a condition necessary for maintaining normal blood pressure
c Act with epinephrine and norepinephrine to produce an anti-inflammatory effect, to bring about normal recovery from inflammations of various kinds
d Produce anti-immunity, antiallergy effect; glucocorticoids bring about a decrease in the number of lymphocytes and plasma cells and therefore a decrease in the amount of antibodies formed
e Glucocorticoid secretion quickly increases when body is thrown into condition of stress; high blood concentration of glucocorticoids, in turn, brings about many other stress responses (see Fig. 8-9)
4 Mineralocorticoids—tend to increase blood sodium and decrease body potassium concentrations by acclerating kidney tubule reabsorption of sodium and excretion of potassium

B Adrenal medulla
 1 Names of hormones—epinephrine (adrenaline) and norepinephrine
 2 Functions of hormones—epinephrine and norepinephrine help body resist stress by intensifying and prolonging effects of sympathetic stimulation; increased epinephrine secretion first endocrine response to stress

Islands of Langerhans

A Names of hormones
 1 Glucagon—secreted by alpha cells
 2 Insulin—secreted by beta cells
B Functions of hormones
 1 Glucagon tends to increase blood glucose by accelerating liver glycogenolysis (conversion of glycogen to glucose)
 2 Insulin tends to decrease blood glucose by accelerating movement of glucose out of blood into cells, which tends to increase glucose metabolism by cells

Female sex glands

Ovaries contain two kinds of cells that secrete hormones—cells of ovarian follicles and of corpus luteum; see Chapter 11

Male sex glands

Interstitial cells of testes secrete male hormone testosterone, which promotes development of male sex organs and male secondary sex characteristics

Thymus

A Name of hormone—thymosin
B Function of hormone—thymosin plays an important role in the development and functioning of the body's immune system

Placenta

A Name of hormones—chorionic gonadotropins
B Function of hormones—maintain the corpus luteum

Endocrine diseases

A Owing to abnormal amounts of growth hormone
 1 Giantism—hypersecretion of growth hormone during early years of life
 2 Acromegaly—hypersecretion of growth hormone in adult years
B Owing to abnormal amounts of thyroid hormone
 1 Goiter—hypersecretion of thyroid hormone

2 Cretinism—hyposecretion of thyroid hormone during early years of life
 3 Myxedema—hyposecretion of thyroid hormone in adult years
C Owing to hypersecretion of glucocorticoids—Cushing's syndrome
D Owing to hyposecretion of insulin—diabetes mellitus
E Owing to hyposecretion of ADH—diabetes insipidus

New words

acromegaly	goiter
antidiuresis	hypercalcemia
catecholamines	hyperglycemia
corticoids	hypoglycemia
cretinism	luteinization
Cushing's syndrome	melanin
diabetes insipidus	mineralocorticoids
diabetes mellitus	myxedema
diuresis	prostaglandins
giantism	stress
glucocorticoids	target organ cell
gluconeogenesis	tetany
glycogenolysis	

Review questions

1 What endocrine glands are located in the following parts of the body:
  abdominal cavity   neck
  cranial cavity   pelvic cavity
  mediastinum
2 What endocrine gland is known as the "master gland"? Why?
3 How are the prostaglandins similar or dissimilar to regular hormones? Name three classes of prostaglandins.
4 What endocrine gland secretes each of the following:
  ACTH   epinephrine
  aldosterone   growth hormone
  calcitonin   insulin
  chorionic   oxytocin
  gonadotropins   progesterone

5 Many changes occur in the body when it is in a condition of stress; for example, after a person has had major surgery. Name two endocrine glands that greatly increase their se-

University of Glamorgan
Learning Resources Centre - Glyntaff

Self Issue Receipt (GT2)

Title: Health assessment
ID: 7312987549
Due: 23:59 Sunday, 30 September 2012

Total items: 1
14/06/2012 16:51
Hold requests: 0
Ready for pickup: 0

Thank you for using the Self-Service system
Diolch yn fawr

7 = D.
21 = Set D
22 = Sand

D Assessment

Health ass

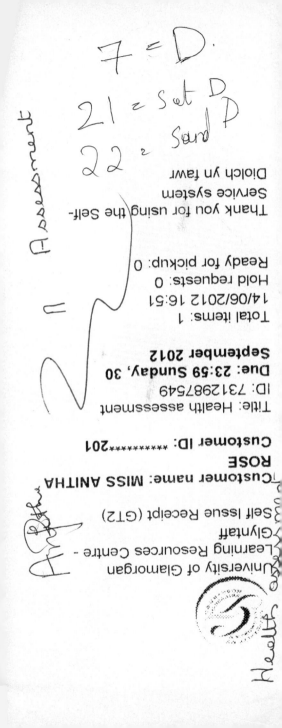

cretion of hormones in times of stress. Name the hormones they secrete.

6 Metabolism changes when the body is in a condition of stress. How does the metabolism of proteins, of fats, and of carbohydrates change, and what hormones cause the changes?

7 What hormones, with a high concentration present in the blood, tend to make us less immune to infectious diseases?

8 What hormone is called the "water-retaining hormone" because it decreases the amount of urine formed?

9 What hormone tends to cause potassium loss from the body through the urine?

10 What hormone is called the "salt-retaining hormone" because it makes the kidneys reabsorb sodium into the blood more rapidly, so that less sodium is lost in the urine?

11 Name two or more hormones that tend to increase blood glucose.

12 What hormone speeds up the rate of catabolism—that is, makes you burn up your foods faster?

13 What is the main function of each of the following:

ACTH	insulin
ADH	parathyroid hormone
calcitonin	thyroid hormone
epinephrine	testosterone

14 Which hormone is called the ovulating hormone?

15 Does acromegaly result from deficient or excess secretion of what hormone?

16 Does cretinism result from deficient or excess secretion of what hormone?

17 Does Cushing's syndrome result from deficient or excess secretion of what hormone?

18 Does diabetes insipidus result from deficient or excess secretion of ADH or insulin?

19 Does diabetes mellitus result from deficient or excess secretion of ADH or insulin?

20 What endocrine disorder produces giantism?

21 What hormone tends to decrease blood calcium concentration? What effect does this have on bone breakdown?

22 What hormone produces opposite effects from the hormone referred to in question 21?

23 What hormone or hormones prepare the body for strenuous activity—for "fight or flight" in other words?

24 What hormone is important in the development and functioning of the immune system?

25 Name the hormone found in the urine of pregnant women.

Systems that provide transportation and immunity

chapter 9

Blood

Chapters 9 and 10 deal with transportation, one of the body's most important functions. Have you ever thought what would happen if all the transportation ceased in your city or town? How soon would you have no food to eat and how soon would rubbish and waste pile up, for instance? Stretch your imagination just a little and you can visualize many disastrous results. Lack of food transportation alone would threaten every individual's life. Similarly, lack of food transportation to cells—the individuals of the body—threatens their lives. The system that provides this vital transportation service is the circulatory system. In this chapter we shall discuss the complex transportation fluid—blood. It is blood that performs the vital pickup and delivery services so necessary for survival of the body's cells. The heart, blood vessels, lymphatic vessels, and lymph nodes will be discussed in Chapter 10.

Blood structure and functions

Cells

Blood is a fluid that has many kinds of chemicals dissolved in it and millions upon millions of cells floating in it. The liquid part of blood is called *plasma*.

There are three main types and several subtypes of blood cells as follows:
1. Red blood cells or erythrocytes
2. White blood cells or leukocytes

a. Granular leukocytes (have granules in their cytoplasm)
   (1) Neutrophils
   (2) Eosinophils
   (3) Basophils
b. Nongranular leukocytes (do not have granules in their cytoplasm)
   (1) Lymphocytes
   (2) Monocytes

Examine Fig. 9-1 to see what each of these different kinds of blood cells looks like under the microscope.

The number of blood cells in the body is hard to believe. For instance, 5,000,000 red blood cells, 7,500 white blood cells, and 300,000 platelets in 1 cubic millimeter of blood (a drop only about $1/25$ of an inch long and wide and high) would be considered normal red blood cell, white blood cell, and platelet counts. The term "formed elements" is often used to describe these nonfluid elements of the blood. Since both red and white blood cells are continually being destroyed, the body has to continually make new ones to take their place at a really staggering rate—a few million red blood cells alone each second.

Two kinds of connective tissue—*myeloid tissue* and *lymphatic tissue*—make blood cells for the body. A better-known name for myeloid tissue is red bone marrow. In the adult body it is present chiefly in the sternum, ribs, hipbones, and cranial bones. A few other bones such as the vertebrae and clavicles also contain small amounts of this valuable substance. Red bone marrow forms all types of blood cells except some of the lymphocytes and monocytes. Most of these are formed by lymphatic tissue located chiefly in the lymph nodes, thymus, and spleen.

Sometimes for one reason or another the blood-forming tissues cannot maintain normal numbers of blood cells. If the number of red blood cells drops below normal, the resulting condition is called *anemia*. If the bone marrow produces an excess of red blood cells, the result is a condition called *polycythemia*. A common laboratory test called the *hematocrit* can tell a physician a great deal about the volume of red cells in a blood sample. If whole blood is placed in a special hematocrit tube and then "spun down" in a centrifuge, the heavier formed elements will quickly settle to the bottom of the tube. During the hematocrit procedure, the red blood cells are forced to the bottom of the tube first. The white blood cells and platelets then settle out in a layer called the *buffy coat*. In Fig. 9-2 the buffy coat can be seen between the packed red blood cells on the bottom of the hematocrit tube and the liquid layer of plasma above. Normally, about 45% of the blood volume is red cells. In anemia the percent of red cells will drop and in polycythemia it will increase dramatically (Fig. 9-2).

Leukopenia, or an abnormally low white blood cell count, occurs only occasionally. However, an abnormally high white blood cell count, *leukocytosis*, is quite common, since it almost always accompanies infec-

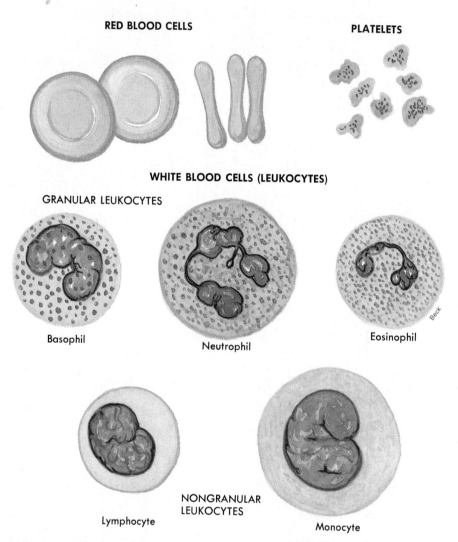

RED BLOOD CELLS

PLATELETS

WHITE BLOOD CELLS (LEUKOCYTES)

GRANULAR LEUKOCYTES

Basophil

Neutrophil

Eosinophil

NONGRANULAR
LEUKOCYTES

Lymphocyte

Monocyte

Fig. 9-1 Human blood cells. There are approximately 30 trillion blood cells in the adult. Each cubic millimeter of blood contains from $4\frac{1}{2}$ to $5\frac{1}{2}$ million red blood cells, 7,500 white blood cells, and an average of 300,000 platelets.

tions. There is also a malignant disease, *leukemia,* in which the number of white blood cells increases tremendously. You may have heard of this disease as "blood cancer." The buffy coat is thicker and more noticeable in the hematocrit of blood taken from leukemia patients. Why?

Red blood cell functions. Red blood cells perform several important functions. They transport oxygen to all the other cells in the body. The *hemoglobin* in them unites with oxygen to form oxyhemoglobin. If hemoglobin falls below the normal level, as it does in anemia, it starts an unhealthy chain reaction: less hemoglobin—less oxygen transported to cells—slower catabolism by cells—less energy supplied to cells—decreased cellular functions. If you under-

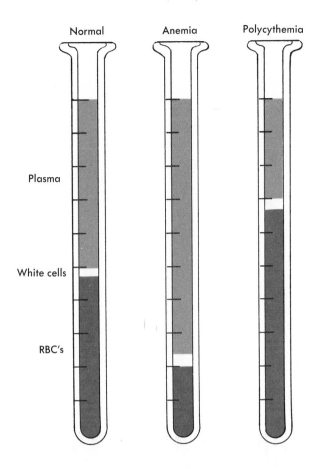

Normal   Anemia   Polycythemia

Plasma

White cells

RBC's

Fig. 9-2 Hematocrit tubes showing normal blood, anemia, and polycythemia. Note the buffy coat located between the packed red cells and the plasma. **A,** Normal percent RBC; **B,** Anemia (low percent RBC); **C,** Polycythemia (high percent RBC).

White blood cell functions. White blood cells carry on a function perhaps slightly less vital than that of red cells but one, nevertheless, that often saves our lives. They defend the body from microorganisms that have succeeded in invading the tissues or bloodstream. Neutrophils and monocytes, for example, engulf microbes. They actually take them into their own cell bodies and digest them. The process is called *phagocytosis,* and the cells that carry on this process are called *phagocytes.* Most numerous of the phagocytes are the neutrophils. However, in addition to neutrophils and monocytes, the body also has other phagocytes that are not white blood cells. They are connective tissue cells of the type classed as reticuloendothelial cells. The liver, spleen, lymph nodes, and bone marrow contain large numbers of reticuloendothelial cells.

White blood cells of the type called lymphocytes also help protect us against infections, but they do it by a process different from phagocytosis. Lymphocytes function in the immune mechanism, the complex process that makes us immune to infectious diseases. The immune mechanism starts to operate, for example, when microbes invade the body. In some way or another their presence acts to stimulate lymphocytes to start multiplying and to become transformed into plasma cells. Presumably, each kind of microbe can stimulate only one specific kind of lymphocyte to multiply and form one kind of plasma cell that makes a specific antibody. The antibody made is specific for destroying the particular microbe that invaded the body in the first

stand this relation between hemoglobin and energy, you can guess correctly what an anemic person's chief complaint will probably be—that he feels "so tired all the time." Red blood cells perform one other essential function; besides transporting oxygen, they also transport carbon dioxide.

As you can see in Fig. 9-1, red blood cells have a most unusual shape. The cell is caved in on both sides so that each one has a thin center and thicker edges. Because of this unique shape and because of the large numbers of red cells, their total surface area is enormous! It provides an area larger than a football field for the exchange of oxygen and carbon dioxide between blood and the body's cells.

place and set the immune mechanism in op-eration. Details of the immune system will be discussed in Chapter 11.

Platelet function (blood clotting). Platelets, the third main type of blood cell, play an essential part in blood clotting. Your life might someday be saved just because your blood can clot. A clot plugs up torn or cut vessels and so stops bleeding that otherwise might prove fatal. The story of how blood clots is the story of a chain of rapid-fire re-actions. The first step in the chain is some kind of an injury to a blood vessel that makes a rough spot in its lining. (Normally the lining of blood vessels is extremely smooth.) Almost immediately some of the blood platelets break up as they flow over the rough spot in the vessel's lining and re-lease a substance into the blood, which leads to the formation of other substances called *platelet factors*. Platelets become "sticky" at the point of injury and soon ac-cumulate near the opening in the cut blood vessel. As platelet numbers increase, so does the release of platelet factors. Soon the next step in the blood clotting mechanism is triggered. Platelet factors combine with prothrombin (a protein present in normal blood), calcium, and other substances to form *thrombin*. Then in the last step throm-bin reacts with fibrinogen (another protein present in normal blood) to change it to a gel called fibrin. Under the microscope fi-brin looks like a tangle of fine threads with red blood cells caught in the tangle. Figs. 9-3 and 9-4 illustrate the steps in the blood clotting mechanism.

The clotting mechanism just described contains clues for ways to stop bleeding by speeding up blood clotting. For example, you might simply apply gauze to a bleeding surface. Its slight roughness would cause more platelets to break down and release more platelet factors. The additional plate-let factors would then make the blood clot more quickly.

Physicians frequently prescribe vitamin K before surgery. Why? The reason is to make sure that the patient's blood will clot fast enough to prevent hemorrhage. Fig. 9-5 shows the somewhat roundabout way in which vitamin K acts to hasten blood clot-ting.

Unfortunately, clots sometimes form in uncut blood vessels of the heart, brain, lungs, or some other organ—a dreaded thing, because clots may produce sudden death by shutting off the blood supply to a vital organ. When a clot stays in the place where it formed, it is called a *thrombus*,

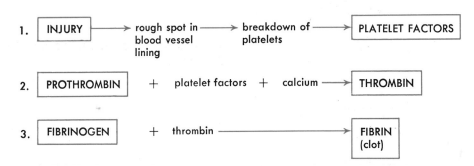

Fig. 9-3 Diagram showing the main steps in blood clotting—a process far more complex than is indicated here.

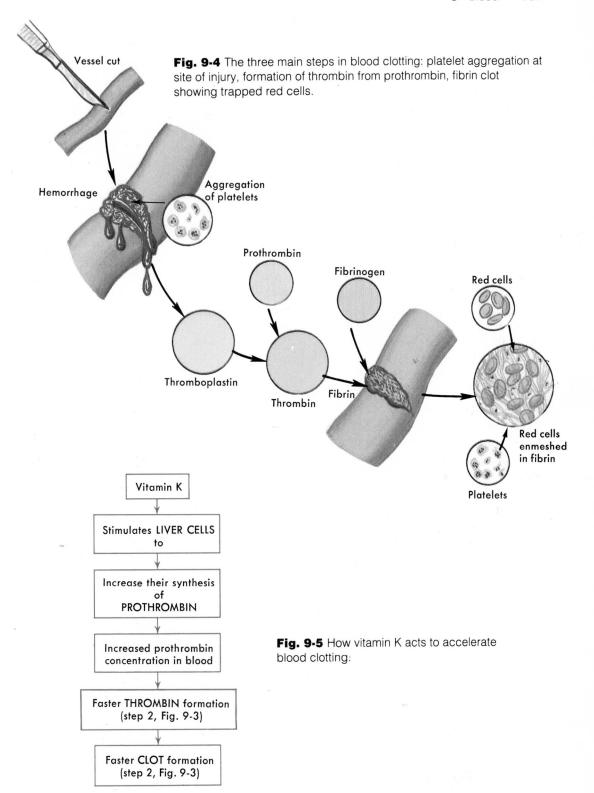

Fig. 9-4 The three main steps in blood clotting: platelet aggregation at site of injury, formation of thrombin from prothrombin, fibrin clot showing trapped red cells.

Vessel cut

Hemorrhage

Aggregation of platelets

Prothrombin

Fibrinogen

Red cells

Thromboplastin

Thrombin

Fibrin

Red cells enmeshed in fibrin

Platelets

Vitamin K

↓

Stimulates LIVER CELLS to

↓

Increase their synthesis of PROTHROMBIN

↓

Increased prothrombin concentration in blood

↓

Faster THROMBIN formation (step 2, Fig. 9-3)

↓

Faster CLOT formation (step 2, Fig. 9-3)

Fig. 9-5 How vitamin K acts to accelerate blood clotting.

and the condition is spoken of as *thrombosis*. If part of the clot dislodges and circulates through the bloodstream, the dislodged part is then called an *embolus* and the condition is called an *embolism*. Suppose that your doctor told you that you had a clot in one of your coronary arteries. Which diagnosis would he make—coronary thrombosis or coronary embolism—if he thought that the clot had formed originally in the coronary artery as a result of the accumulation of fatty material in the vessel wall? Doctors now have some drugs that they can use to help prevent thrombosis and embolism. Dicumarol is one. It blocks the stimulating effect of vitamin K on the liver, and consequently the liver cells make less prothrombin. The blood prothrombin content soon falls low enough to prevent clotting.

Blood plasma

Blood plasma is the liquid part of the blood, or blood minus its cells. It consists of water with many substances dissolved in it. All of the chemicals needed by cells to stay alive—food, oxygen, and salts, for example—have to be brought to them by the blood. Food and salts are dissolved in plasma. So, too, is a small amount of oxygen. (Most of the oxygen in the blood is carried in the red blood cells as oxyhemoglobin.) Wastes that cells must get rid of are dissolved in plasma and transported to the excretory organs. And, finally, the hormones that help control our cells' activities and the antibodies that help protect us against microorganisms are dissolved in plasma.

Many people seem curious about how much blood they have. The amount depends on how big they are and whether they are male or female. A big person has more blood than a small person and a man has more blood than a woman. But as a general

rule, most adults probably have between 4 and 6 quarts (4 and 6 liters) of blood. It normally accounts for about 7% to 9% of the total body weight.

The volume of the plasma part of blood is usually a little more than half the volume of whole blood. Blood cells make up the remaining part of whole blood's volume. Examples of normal volumes are the following: plasma volume—2.6 liters; blood cell volume—2.4 liters; total blood volume—5 liters.

If you read many advertisements or watch many television commercials, you may think that almost everyone has "acid blood" at some time or other. Nothing could be farther from the truth. Blood is alkaline; it rarely reaches even the neutral point. If the alkalinity of your blood decreases toward neutral, you are a very sick person; in fact, you have what is called *acidosis*. But even in this condition, blood almost never becomes the least bit acid; it just becomes less alkaline than normal.

Blood types (or blood groups)

Blood types are identified by the presence on the red blood cells of certain antigens. An *antigen* is a substance that can stimulate the body to make antibodies. Almost all substances that act as antigens are foreign proteins. In other words, they are not the body's own natural proteins but are proteins that have entered the body from the outside—by injection or transfusion or some other method.

The word "antibody" can be defined in terms of what causes its formation or in terms of how it functions. Defined the first way, an *antibody* is a substance made by the body in response to stimulation by an antigen. Defined according to its functions, an antibody is a substance that reacts with the antigen that stimulated its formation. Many antibodies react with their antigens

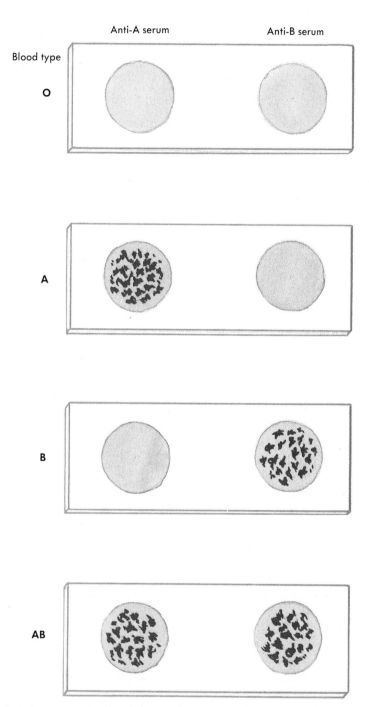

Fig. 9-6 Appearance of agglutination or clumping tests used to type blood. Blood type O red cells do not clump when mixed with either anti-A or anti-B serum; blood type A red cells clump when mixed with anti-A serum; and blood type B red cells clump with anti-B serum. Red cells in type AB blood are agglutinated (clumped) by both anti-A and anti-B serums.

to clump (or agglutinate) them. In other words, they cause their antigens to stick together in little clusters.

Every person's blood belongs to one of the following four blood types: type A, type B, type AB, or type O. Suppose that you have type A blood (as do about 41% of Americans). The letter *A* stands for a certain type of antigen (a protein) present in the cytoplasmic membrane of your red blood cells when you were born. Because you were born with type A antigen, your body does not form antibodies to react with it. In other words, your blood plasma contains no anti-A antibodies. It does, however, contain anti-B antibodies. For some unknown reason these antibodies are present naturally in type A blood plasma. The body did not form them in response to the presence of their antigen. In summary, then, in type A blood the red blood cells contain type A antigen and the plasma contains anti-B antibodies.

In type AB blood, as its name indicates, the red blood cells contain both type A and type B antigens and the plasma contains neither anti-A nor anti-B antibodies. The opposite is true of type O blood—its red blood cells contain neither type A nor type B antigens and its plasma contains both anti-A and anti-B antibodies.

Harmful effects or even death can result from a blood transfusion if the donor's red blood cells become agglutinated by antibodies in the recipient's plasma. If a donor's red blood cells do not contain any A or B antigen, they of course cannot be clumped by anti-A or anti-B antibodies. For this reason the type of blood that contains neither A nor B antigens—namely, type O blood—can be used as donor blood without the danger of anti-A or anti-B antibodies clumping its red blood cells. Type O blood is therefore called *universal donor* blood. *Universal recipient* blood is type AB; it contains neither anti-A nor anti-B antibodies in its plasma. Therefore it cannot clump any donor's red blood cells containing A or B antigens.

In recent years the expression *Rh-positive* blood has become a familiar one. It means that the red blood cells of this type blood contain an antigen called the Rh factor. If, for example, a person has type AB, Rh-positive blood, his red blood cells contain type A antigen, type B antigen, and the Rh factor.

In *Rh-negative* blood the red blood cells do not contain the Rh factor. Plasma never naturally contains anti-Rh antibodies. But if Rh-positive blood cells are introduced into an Rh-negative person's body, anti-Rh antibodies soon appear in his blood plasma. In this fact lies the danger for a baby born to an Rh-negative mother and Rh-positive father. If the baby is Rh-positive, the Rh factor on its red blood cells may stimulate the mother's body to form anti-Rh antibodies. Then, if she later carries another Rh-positive fetus, it may develop a disease called erythroblastosis fetalis, caused by the mother's Rh antibodies reacting with the baby's Rh-positive cells.

Outline summary
Blood structure and functions
A Cells
  1 Kinds
    a Red blood cells (erythrocytes)
    b White blood cells (leukocytes)
      (1) Granular leukocytes—neutrophils, eosinophils, basophils
      (2) Nongranular leukocytes—lymphocytes, monocytes
  2 Numbers
    a Red blood cells—4½ to 5 million per cubic millimeter of blood
    b White blood cells—5,000 to 9,000 per cubic millimeter of blood
    c. Platelets—300,000 per cubic millimeter of blood
  3 Formation—red bone marrow (myeloid tis-

sue) forms all blood cells except some lymphocytes and monocytes, which are formed by lymphatic tissue in lymph nodes, thymus, and spleen
4 Red blood cell functions—transport oxygen and carbon dioxide
5 White blood cell functions—neutrophils and monocytes carry on phagocytosis; lymphocytes function to produce immunity
6 Platelet functions—Blood clotting (summarized in Figs. 9-3 and 9-4)
B Blood plasma
1 Definition—blood minus its cells
2 Composition—water containing many dissolved substances, for example, foods, salts, hormones
3 Amount of blood—varies with size and sex; 4 to 6 liters about average; about 7% to 9% of body weight
4 Reaction—slightly alkaline
C Blood types (or blood groups) (see Fig. 9-6)
1 Type A blood—type A antigens in red cells; anti-B type antibodies in plasma
2 Type B blood—type B antigens in red cells; anti-A type antibodies in plasma
3 Type AB blood—type A and type B antigens in red cells; no anti-A or anti-B antibodies in plasma; therefore type Ab blood is called universal recipient blood
4 Type O blood—no type A or type B antigens in red cells; therefore type O blood is called universal donor blood; both anti-A and anti-B antibodies in plasma
5 Rh-positive blood—Rh factor antigen present in red blood cells
6 Rh-negative blood—no Rh factor present in red blood cells; no anti-Rh antibodies present naturally in plasma; anti-Rh antibodies, however, appear in plasma of Rh-negative person if Rh-positive blood cells have been introduced into his body

Review questions

1 What is the normal number of red blood cells per cubic millimeter of blood? white blood cells? platelets?
2 Name the granular leukocytes.
3 What two kinds of connective tissue make blood cells for the body?
4 Suppose that your doctor told you that your "red count was 3 million." What does "red count" mean? Might the doctor say that you had any of the following conditions—anemia, leukocytosis, leukopenia, polycythemia—with a red blood count of this amount? If so, which one?
5 What does the hematocrit blood test measure? What is the normal value for this test?
6 If you had appendicitis or some other acute infection, would your white blood cell count be more likely to be 2,000, 7,000, or 15,000? Give a reason for your answer.
7 Your circulatory system is the transportation system of your body. Mention some of the substances it transports and tell whether each is carried in blood cells or in the blood plasma.
8 What organs of the body contain large numbers of reticuloendothelial cells? What role do these cells play in the circulatory system?
9 Briefly explain what happens when blood clots, including what makes it start to clot.
10 You hear that a friend has a "coronary thrombosis." What does this mean to you?
11 Define the terms "antibody" and "antigen" as they apply to blood typing.
12 Identify the four blood types.
13 What is meant by the terms "universal donor" and "universal recipient"?
14 What is the difference between blood type and Rh factor?
15 Why would a physician prescribe vitamin K before surgery for a patient with a history of bleeding problems?

New words

anemia	fibrinogen	prothrombin
antibodies	hematocrit	serum
antigens	leukemia	thrombin
buffy coat	leukocytosis	thrombosis
embolism	leukopenia	thrombus
fibrin	plasma	

chapter 10

Cardiovascular and lymphatic systems

Billions of cells make up our bodies, and each cell in order to survive must have numerous substances transported to and from it. The system that supplies our cells' transportation needs is the circulatory system—an appropriate name because it consists of a pump, the heart, which keeps a fluid, the blood, moving around and around through a closed circle of blood vessels. Not only must the blood be kept moving through the vessels by the heart's pumping action, but also the amount of blood flowing to specific regions must be varied according to their degree of activity. When a structure's activity increases, the volume of blood flow to the structure also increases. For example, during exercise, skeletal muscle activity increases and so, too, does the volume of blood flowing to muscles increase; following a meal, digestive organ activity increases and blood flow to digestive organs also increases. This chapter will discuss the heart, blood vessels, blood flow regulation, and a special part of the circulatory system called the lymphatic system.

Heart

Location, size, and position

No one needs to be told where his heart is or what it does. Everyone knows that the heart is in the chest, that it beats night and day to keep the blood flowing, and that if it stops, life stops.

Most of us probably think of the heart as located on the left side of the body. As you can see in Fig. 10-1, the heart is located between the lungs in the lower portion of the mediastinum. Draw an imaginary line through the middle of the trachea in Fig. 10-1 and continue the line down through the thoracic cavity to divide it into right and left halves. Note that about two thirds of the mass of the heart is to the left of this line and one third to the right. The heart is often described as a triangular organ, shaped and sized roughly like a closed fist. In Fig. 10-1 you can see that the *apex* (blunt point of the lower edge of the heart) lies on the diaphragm, pointing toward the left. To count the apical beat, one must place a stethoscope directly over the apex, that is, in the space between the fifth and sixth ribs on a line with the midpoint of the left clavicle.

The heart is positioned in the thoracic cavity between the sternum in front and the bodies of the thoracic vertebrae behind. Because of this placement, it can be compressed or squeezed by application of pressure to the lower portion of the body of the sternum using the heel of the hand. Rhythmic compression of the heart in this way can maintain blood flow in cases of cardiac arrest and, if combined with effective artificial respiration, the resulting procedure, called *cardiopulmonary resuscitation* (CPR), can be lifesaving.

Internal structure and covering sac

If you cut open a heart, you can see many of its main structural features (Fig. 10-2). It is hollow, not solid. A partition (the septum) divides it into right and left sides. It has four cavities inside: a small upper cavity *(atrium)* and a larger lower cavity *(ventricle)* on each side. Cardiac muscle tissue composes the wall of the heart; it is usually referred to as the *myocardium.* Myocarditis therefore means inflammation of the heart muscle. A very smooth tissue, *endocardium*, lines the cavity of the heart. Endocarditis, of course, means inflammation of the heart lining. This condition can cause rough spots to develop in the endocardium, which may lead to thrombosis. (Look back at Figs. 9-3 and 9-4 to find out why.) Rough spots, such as injuries to blood vessel walls, cause release of platelet factors. The result is often the formation of a fatal blood clot.

The heart has a covering as well as a lining. Its covering, the *pericardium*, consists of two layers of fibrous tissue with a small space in between. The inner layer of the pericardium covers the heart like an apple skin covers an apple, but the outer layer fits around the heart like a loose-fitting sack, allowing enough room for the heart to beat in it. These two pericardial layers slip

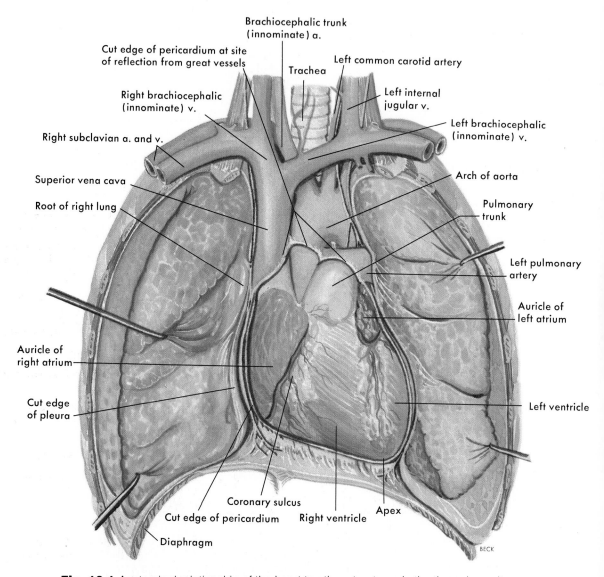

Fig. 10-1 Anatomical relationship of the heart to other structures in the thoracic cavity.

against each other without friction when the heart beats because they are moist, not dry surfaces. (Have you ever walked carefully on a wet floor? If so, you were consciously or unconsciously using your knowledge of the principle that moist surfaces are slippery.) A thin film of pericardial fluid furnishes the lubricating moistness between the heart and its enveloping pericardial sac.

Four valves keep blood flowing through the heart in one direction (Fig. 10-2). The *tricuspid valve*, at the opening between the

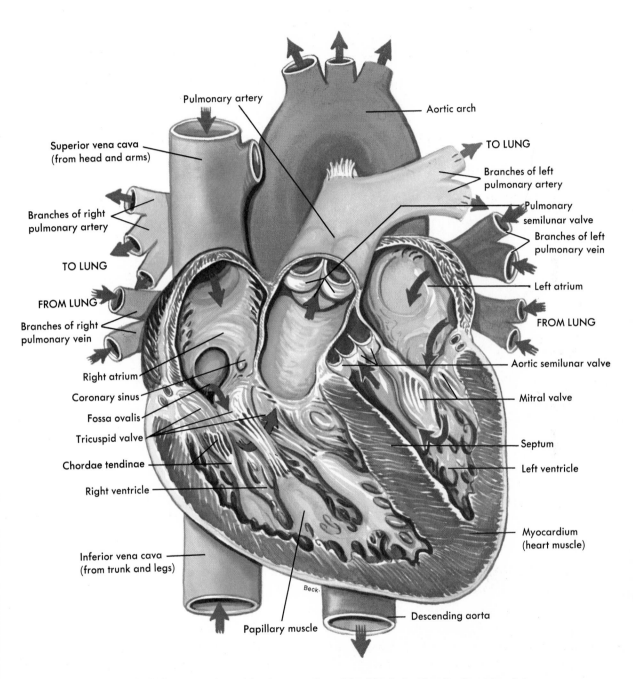

Pulmonary artery

Aortic arch

Superior vena cava
(from head and arms)

TO LUNG

Branches of left
pulmonary artery

Pulmonary
semilunar valve

Branches of right
pulmonary artery

Branches of left
pulmonary vein

TO LUNG

Left atrium

FROM LUNG

FROM LUNG

Branches of right
pulmonary vein

Aortic semilunar valve

Right atrium

Coronary sinus

Mitral valve

Fossa ovalis

Tricuspid valve

Septum

Chordae tendinae

Left ventricle

Right ventricle

Myocardium
(heart muscle)

Inferior vena cava
(from trunk and legs)

Beck.

Descending aorta

Papillary muscle

Fig. 10-2 Cutaway view of the front section of the heart showing the four chambers, the valves, the openings, and the major vessels. Arrows indicate the direction of blood flow; the blue arrows represent unoxygenated blood and the red arrows oxygenated blood.

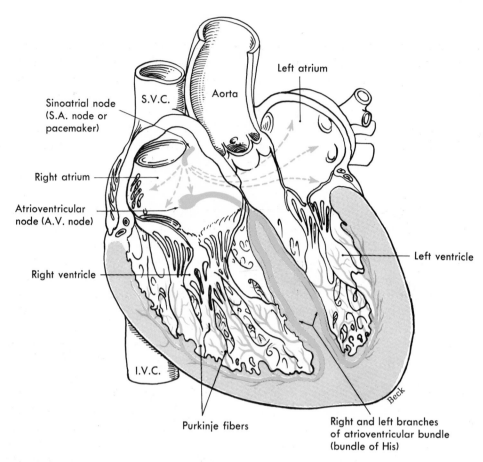

Left atrium

Aorta

S.V.C.

Sinoatrial node
(S.A. node or
pacemaker)

Right atrium

Atrioventricular
node (A.V. node)

Right ventricle

Left ventricle

I.V.C.

Beck

Purkinje fibers

Right and left branches
of atrioventricular bundle
(bundle of His)

Fig. 10-3 The conduction system of the heart. The sinoatrial node in the wall of the right atrium sets the basic pace of the heart's rhythm, so it is called the "pacemaker."

right atrium and the right ventricle, lets blood flow from the atrium into the ventricle but prevents it from flowing in the opposite direction. The *mitral* or *bicuspid valve*, at the opening between the left atrium and the left ventricle, allows blood flow from the atrium into the ventricle but prevents flow from the ventricle back up into the atrium. The *pulmonary semilunar valve* at the beginning of the pulmonary artery allows blood to flow out of the right ventricle but prevents it from backflowing into the ventricle. The *aortic semilunar valve* at the beginning of the aorta allows blood to flow out of the left ventricle up into the

aorta but prevents backflow into this ventricle.

Conduction system

Embedded in the wall of the heart are four structures that conduct impulses through the heart muscle to cause first the atria and then the ventricles to contract. The names of the structures that make up this conduction system of the heart are the *sinoatrial node* (SA node, or the pacemaker of the heart), *atrioventricular node* (AV node), the *bundle of His* (or AV bundle), and the *Purkinje* (pur-kin'-jē) *fibers*. Impulse conduction normally starts in the heart's pacemaker, namely, the SA node. From here it spreads, as you can see in Fig. 10-3, in all directions through both atria. This causes the atria to contract. When impulses reach the AV node, it relays them by way of the bundle of His and Purkinje fibers to the ventricles, causing them to contract. Normally therefore a ventricular beat follows each atrial beat. Various diseases, however, can damage the heart's conduction system and thereby disturb the rhythmical beating of the heart. One such disturbance is the condition commonly called *heart block.* Impulses are blocked from getting through to the ventricles, with the result that the heart beats at a much slower rate than normal. A physician may treat heart block by implanting in the heart an *artificial pacemaker,* an electrical device that causes ventricular contractions at a rate fast enough to maintain an adequate circulation of blood.

Blood vessels

Kinds

Arteries, veins, and capillaries—these are the names of the three main kinds of blood vessels. *Arteries* carry blood away from the heart toward capillaries. *Veins* carry blood toward the heart away from capillaries. *Capillaries* carry blood from tiny arteries *(arterioles)* into tiny veins *(venules).* The largest artery in the body is the aorta. The largest veins are the *superior vena cava* and the *inferior vena cava.* The aorta carries blood out of the left ventricle of the heart, and the venae cavae return blood to the right atrium after the blood has circulated through the body.

Structure

Arteries, veins, and capillaries differ in structure. Examine Figs. 10-4 and 10-5.

Three coats or layers are found in both arteries and veins. The outermost layer or coat is called the *tunica adventitia.* Note that muscle tissue is found in the middle layer or *tunica media* of both arteries and veins. It is important to know, however, that the muscle layer is much thicker in arteries than in veins. Why is this important? Because the thicker muscle layer in the artery wall plays such a critical role in maintaining blood pressure and controlling blood distribution in the body. This is smooth muscle and so is controlled by the autonomic nervous system. A thin layer of elastic and white fibrous tissue covers an inner layer of endothelial cells called the *tunica intima* in both arteries and veins. The tunica intima is actually a single layer of endothelial cells that lines the inner surface of these vessels. When a surgeon cuts into the body, only arteries, arterioles, veins, and venules can be seen. Capillaries cannot be seen because they are microscopic vessels. The most important structural feature of capillaries is their extreme thinness—only one layer of flat, endothelium-like cells composes the capillary membrane. Instead of three layers or coats the capillary wall is composed of only one—the tunica intima. Substances such as glucose, oxygen, and wastes can quickly pass through it on their way to or from cells.

Functions

Arteries, veins, and capillaries serve different functions. Arteries and arterioles distribute blood from the heart to capillaries in all parts of the body. In addition, arterioles, by constricting or dilating, help maintain arterial blood pressure at a normal level. Venules and veins collect blood from capillaries and return it to the heart. They also serve as blood reservoirs, since they can expand to hold a larger volume of blood or constrict to hold a much smaller

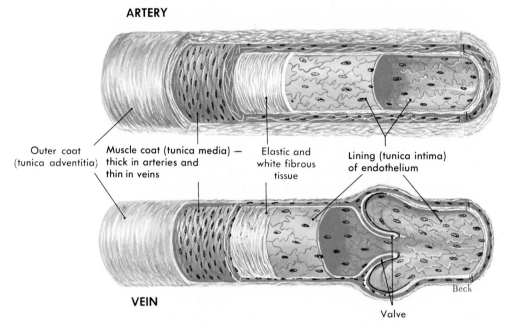

ARTERY

Outer coat
(tunica adventitia) | Muscle coat (tunica media) —
thick in arteries and
thin in veins | Elastic and
white fibrous
tissue | Lining (tunica intima)
of endothelium

VEIN

Valve

Beck

Fig. 10-4 Schematic drawings of an artery and vein showing comparative thickness of the three coats: the outer coat (tunica adventitia), the muscle coat (tunica media), and the lining of endothelium (tunica intima). Note that the muscle and outer coats are much thinner in veins than in arteries and that veins have valves.

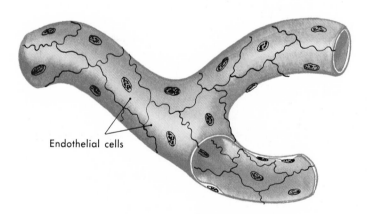

Endothelial cells

Fig. 10-5 The walls of capillaries consist of only a single layer of endothelial cells. These thin, flattened cells permit the rapid movement of substances between blood and interstitial fluid. Note that capillaries have no smooth muscle layer, elastic fibers, or surrounding coats.

amount. Capillaries function as exchange vessels. For example, glucose and oxygen move out of the blood in capillaries into interstitial fluid and on into cells. Carbon dioxide and various other substances move in the opposite direction, that is, into the capillary blood from the cells. Fluid also is exchanged between capillary blood and interstitial fluid. (See discussion in Chapter 17.)

Study Fig. 10-6 to find out the names of the main arteries of the body and Figs. 10-7

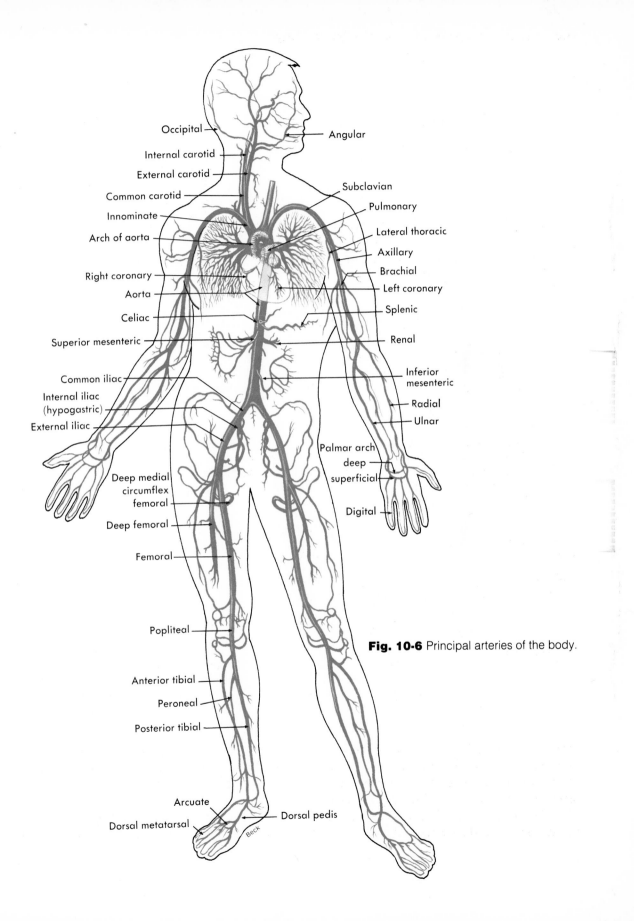

Occipital

Angular

Internal carotid

External carotid

Subclavian

Common carotid

Pulmonary

Innominate

Lateral thoracic

Arch of aorta

Axillary

Brachial

Right coronary

Left coronary

Aorta

Splenic

Celiac

Renal

Superior mesenteric

Inferior mesenteric

Common iliac

Internal iliac (hypogastric)

Radial

External iliac

Ulnar

Palmar arch

deep

superficial

Deep medial circumflex femoral

Digital

Deep femoral

Femoral

Popliteal

Fig. 10-6 Principal arteries of the body.

Anterior tibial

Peroneal

Posterior tibial

Arcuate

Dorsal metatarsal

Dorsal pedis

Beck

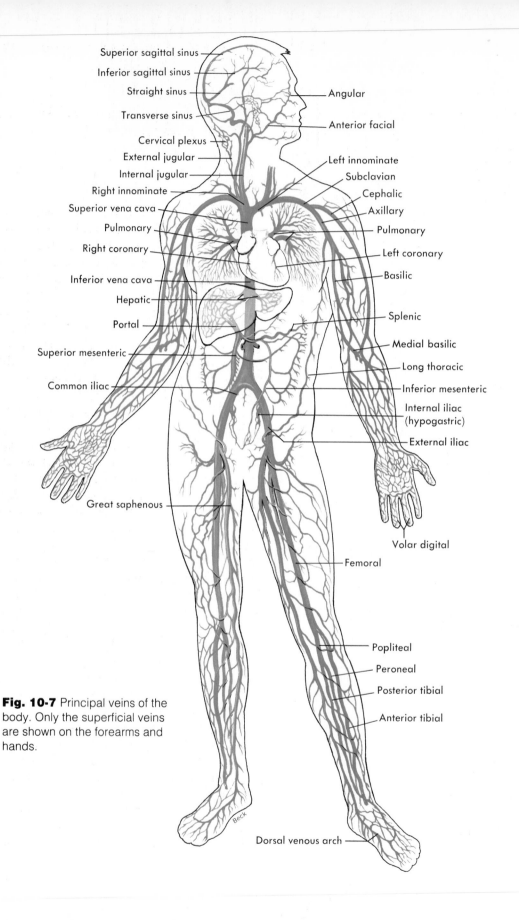

Superior sagittal sinus

Inferior sagittal sinus

Straight sinus

Transverse sinus

Cervical plexus

External jugular

Internal jugular

Right innominate

Superior vena cava

Pulmonary

Right coronary

Inferior vena cava

Hepatic

Portal

Superior mesenteric

Common iliac

Great saphenous

Angular

Anterior facial

Left innominate

Subclavian

Cephalic

Axillary

Pulmonary

Left coronary

Basilic

Splenic

Medial basilic

Long thoracic

Inferior mesenteric

Internal iliac (hypogastric)

External iliac

Volar digital

Femoral

Popliteal

Peroneal

Posterior tibial

Anterior tibial

Dorsal venous arch

Beck

Fig. 10-7 Principal veins of the body. Only the superficial veins are shown on the forearms and hands.

and 10-8 for the names of the main veins. Then see whether you can answer these questions: What is the name of the main artery of the thigh? of the upper arm? of the thumb side of the lower arm? What are neck arteries called? neck veins? Make up

some more questions to quiz yourself about blood vessel names and locations.

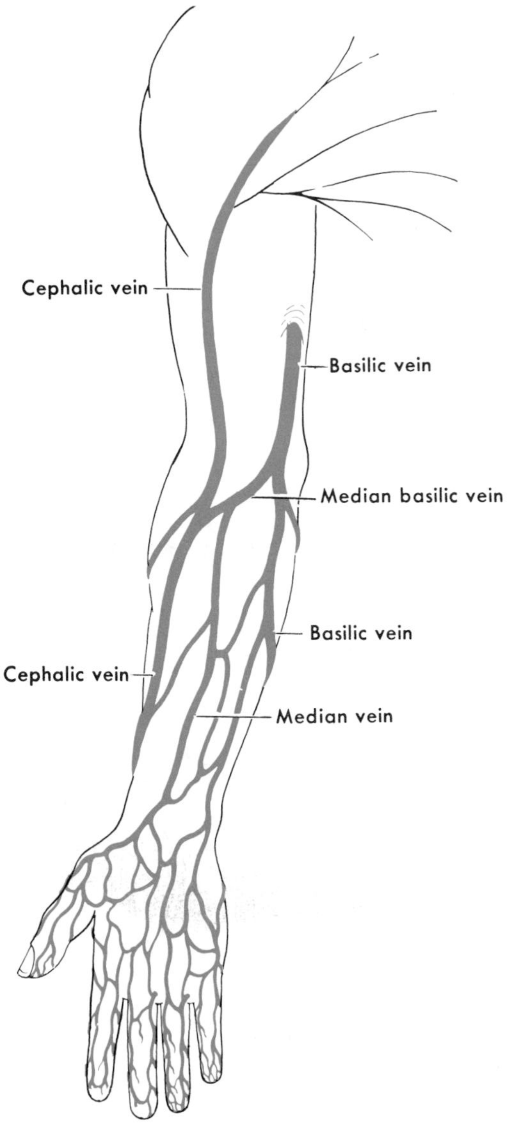

Circulation

The word "circulation" implies that something moves over a circular route and that it moves over this route repeatedly. Applied to blood, circulation means the movement of blood over and over again through vessels that form a circular route. (A circular route begins and ends at the same place but is not necessarily shaped like a circle.) Since the heart pumps the blood and since the blood returns to the right atrium of the heart when it has completed one circuit through the blood vessels, we shall consider that circulation starts in the right atrium. Examine Fig. 10-9 carefully. Imagine that you have injected a drug into a patient's right upper arm. The drug would soon diffuse into the blood in the capillaries of the upper arm. Look again at Fig. 10-9 to find out what kind of vessels the blood would flow into as it left the arm's capillaries. Notice next what two organs the drug-containing blood would have to pass through before it could enter the capillaries of, for instance, the other arm, or in fact of any other organ in the body. First it would have to return to the heart (which side? which chamber?). Next it would have to circulate through the lungs, come back to the heart (which side? which chamber?) and then be pumped by the left ventricle out into the aorta and on into arteries, arterioles, and capillaries of other organs. Make sure that you understand the blood's circulation route by answering the appropriate questions on p. 162 and 163 and checking your answers with Fig. 10-9.

As blood flows through capillaries in the lungs, it changes from venous blood to arterial blood by unloading carbon dioxide

Fig. 10-8 Main superficial veins of the arm.

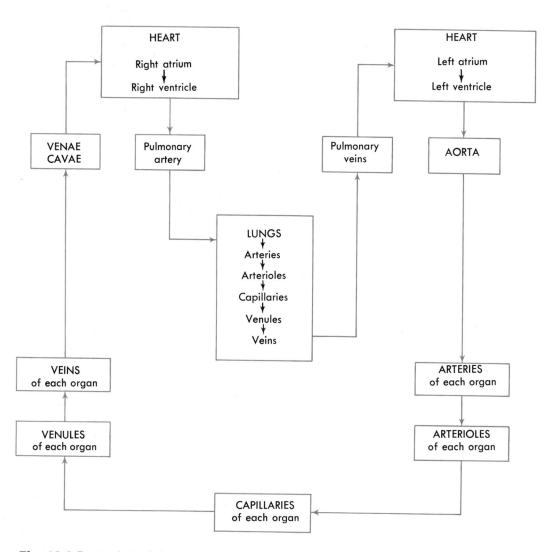

Fig. 10-9 Route of circulating blood.

and picking up oxygen. Its color changes in the process from the deep crimson that identifies venous blood to the bright scarlet of arterial blood. Then as blood flows through tissue capillaries (all capillaries except those in the lungs), it changes back from arterial to venous blood by oxygen leaving the blood to enter cells and by carbon dioxide leaving the cells to enter blood.

Blood flows into the heart muscle itself by way of two small vessels that are surely the most famous of all the blood vessels—the coronary arteries—famous because coronary disease kills so many thousands every year. The coronary arteries are the aorta's first branches. The openings into these small vessels lie behind the flaps of the aortic semilunar valves. In both coronary thrombosis and coronary embolism (pp. 136 and 137), a blood clot occludes, or plugs up, some part of a coronary artery. Blood cannot pass through the occluded vessel and so cannot reach the heart muscle cells it normally supplies. Deprived of oxygen, these cells soon die. In medical terms, myocardial infarction occurs. Myocardial infarction is a common cause of death in middle-aged and old people.

Portal circulation

The term "portal circulation" refers to the route of blood flow through the liver. Veins from the abdominal digestive organs (stomach, pancreas, and intestines) and from the spleen empty into the portal vein. Branches of the portal vein empty into arterioles, which empty into the capillaries of the liver. Blood leaves the liver by way of the hepatic vein, which drains into the inferior vena cava. How does portal circulation differ from circulation through other organs? Does venous blood ordinarily flow into capillaries? See Fig. 10-9 if you are not sure about this. The detour of venous blood through the liver before its return to the heart serves some valuable purposes. For example, when a meal is being absorbed, the blood in the portal vein contains a higher than normal concentration of glucose. Liver cells remove the excess glucose and store it as glycogen. Blood leaving the liver therefore usually has a normal blood glucose concentration. Liver cells also remove and detoxify various poisonous substances that may be present in the blood returning to the intestines.

Fetal circulation

Circulation in a baby before birth is not the same as after birth. This makes good sense because before birth the baby's blood has to get oxygen and food from the mother's blood instead of from its own

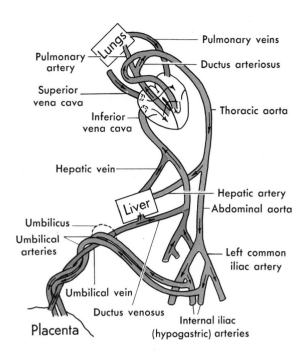

Fig. 10-10 Fetal circulation. Note these six structures that are present in a baby's body at birth, but not in a normal adult body: (1) umbilical arteries (two of them); (2) placenta (afterbirth); (3) umbilical vein; (4) ductus venosus; (5) ductus arteriosus; (6) foramen ovale (the opening—not labeled—between the right and left atria of the heart).

lungs and digestive tract. Therefore relatively little blood needs to circulate through the baby's lungs and digestive organs before it is born, but all of its blood needs to circulate through the placenta (afterbirth) where the exchange of substances with the mother's blood occurs. As you read the next paragraph, refer often to Fig. 10-10.

Two *umbilical arteries* (extensions of the baby's internal iliac arteries) carry blood to the *placenta*, a highly vascular structure attached to the inside of the mother's uterus. As the baby's blood circulates through the placenta, oxygen and foods enter it from the mother's blood while wastes move from the baby's to the mother's blood. After flowing through the placenta, blood returns to the baby's body by way of one *umbilical vein.* (Note in Fig. 10-10 the red color of this vessel to indicate oxygenated blood.) Part of the returning blood circulates through the liver; the rest flows through the *ductus venosus* into the inferior vena cava and on into the right atrium of the baby's heart. Some of the blood then flows to the lungs as it does in the adult body. But some of it flows through two detours that bypass the lungs. One is a hole *(foramen ovale)* in the septum of the heart between the two atria. The other is a small vessel (the *ductus arteriosus)* that leads from the pulmonary artery to the thoracic aorta. After a baby is delivered and the umbilical cord is cut, the umbilical arteries, umbilical vein, and placenta no longer function. Soon the foramen ovale closes, and eventually the ductus venosus and arteriosus become fibrous cords.

Blood pressure

Perhaps a good way to try to understand blood pressure is to try to answer the questions about it of what, where, why, and how. What is blood pressure? Just what the words say—blood pressure is the pressure, or push, of blood.

Where does blood pressure exist? It exists in all blood vessels, but it is highest in the arteries and lowest in the veins. In fact, if we list blood vessels in order according to the amount of blood pressure in them and draw a graph of this, as in Fig. 10-11, the graph looks like a hill, with aortic blood pressure at the top and vena caval pressure at the bottom. This blood pressure "hill" is spoken of as the blood pressure gradient. More precisely, the term "blood pressure gradient" means the difference between two blood pressures. The blood pressure gra-

dient for the entire systemic circulation is the difference between the average, or mean, blood pressure in the aorta and the blood pressure at the termination of the venae cavae where they join the right atrium of the heart. The mean blood pressure in the aorta, given in Fig. 10-11, is 100 millimeters of mercury (mm Hg) and the pressure at the termination of the venae cavae is 0. Therefore, with these typical normal figures, the systemic blood pressure gradient is 100 mm. Hg (100 minus 0).

Why does blood pressure exist? What is its function? The blood pressure gradient is vitally involved in keeping the blood flowing. When a blood pressure gradient is present, blood circulates. Conversely, when a blood pressure gradient is not present, blood does not circulate. For example, suppose that the blood pressure in the arteries were to decrease so that it became equal to the average pressure in arterioles. There would no longer be a blood pressure gradient between arteries and arterioles, and therefore there would no longer be a force to move blood out of arteries into arterioles. Circulation would stop, in other words, and very soon life itself would cease. This is why when arterial blood pressure is observed to be falling rapidly, whether in surgery or elsewhere in a hospital, emergency measures are quickly started to try to reverse this fatal trend.

What we have just said in the preceding paragraph may start you wondering why high blood pressure (meaning, of course, high arterial blood pressure) and low blood pressure are bad for circulation. High blood pressure is considered bad for several reasons. For one thing, if it becomes too high, it may cause the rupture of one or more blood vessels (for example, in the brain, as happens in a stroke). But low blood pressure also can be dangerous. If arterial pressure falls low enough, circulation and life

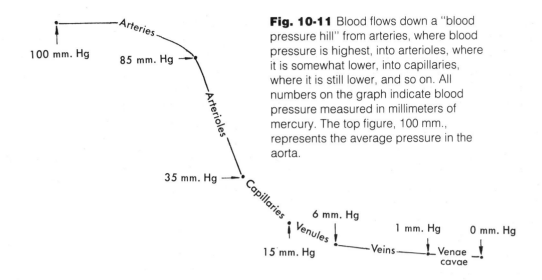

Fig. 10-11 Blood flows down a "blood pressure hill" from arteries, where blood pressure is highest, into arterioles, where it is somewhat lower, into capillaries, where it is still lower, and so on. All numbers on the graph indicate blood pressure measured in millimeters of mercury. The top figure, 100 mm., represents the average pressure in the aorta.

cease. Massive hemorrhage, for instance, kills in this way.

How is blood pressure produced? What causes blood pressure, in other words, and what makes blood pressure change from time to time? The direct cause of blood pressure is the volume of blood in the vessels. The larger the volume of blood in the arteries, for example, the more pressure the blood exerts on the walls of the arteries, or the higher the arterial blood pressure.

Conversely, the less blood in the arteries, the lower the blood pressure tends to be. Hemorrhage demonstrates well this relation between blood volume and blood pressure. In hemorrhage a pronounced loss of blood occurs, and this decrease in the volume of blood causes blood pressure to drop. In fact, the major sign of hemorrhage is a rapidly falling blood pressure.

The volume of blood in the arteries is determined by how much blood the heart pumps into the arteries and how much blood the arterioles drain out of them. (The volume of blood pumped into the arteries is called the *cardiac output*.) The diameter of

the arterioles determines how much blood drains out of arteries into arterioles.

Both the strength and the rate of the heartbeat affect cardiac output and therefore blood pressure. Each time the left ventricle contracts, it squeezes a certain volume of blood (called the stroke volume) into the aorta and on into other arteries. The stronger each contraction is, the more blood it pumps into the aorta and arteries. Conversely, the weaker each contraction is, the less blood it pumps. Suppose that one contraction of the left ventricle pumps 70 milliliters of blood into the aorta, and suppose that the heart beats 70 times a minute. Seventy milliliters times 70 equals 4,900 milliliters. Almost 5 quarts of blood would enter the aorta and arteries every minute. Now suppose that the heartbeat were to become weaker and that each contraction of the left ventricle pumps only 50 milliliters of blood instead of 70 into the aorta. If the heart still contracts 70 times a minute, it will obviously pump much less blood into the aorta—only 3,500 milliliters instead of the more normal 4,900 milliliters per minute. This decrease in the heart's output of blood tends to decrease the volume of blood in the arteries, and the decreased arterial blood volume tends to decrease arterial blood pressure. In summary, the strength of the heartbeat affects blood pressure in this way—a stronger heartbeat tends to increase blood presssure and a weaker beat tends to decrease it.

The rate of the heartbeat may also affect arterial blood pressure. You might reason that when the heart beats faster, more blood would enter the aorta and that therefore the arterial blood volume and blood pressure would increase. This is true only if the stroke volume does not decrease sharply when the heart rate increases. Often, however, when the heart beats faster, each contraction of the left ventricle takes place so rapidly that it squeezes out much less blood than usual into the aorta. For example, suppose that the heart rate speeded up from 70 to 100 times a minute and that at the same time its stroke volume decreased from 70 milliliters to 40 milliliters. Instead of a cardiac output of 70 × 70 or 4,900 milliliters per minute, the cardiac output would have changed to 100 × 40 or 4,000 milliliters per minute. Arterial blood volume would tend to decrease under these conditions and therefore blood pressure would also tend to decrease even though the heart rate had increased. What generalization, then, can we make? We can only say that an increase in the rate of the heartbeat tends to increase blood pressure and a decrease in the rate tends to decrease blood pressure. But whether a change in the heart rate actually produces a similar change in blood pressure depends on whether the strength of the heart's beat also changes and in which direction (as indicated by the stroke volume).

Another factor that we ought to mention in connection with blood pressure is the viscosity of blood, or in plainer language, its stickiness. If blood becomes less viscous than normal, blood pressure decreases. For example, if a person suffers a hemorrhage, fluid will move into his blood from his interstitial fluid. This dilutes his blood and decreases its viscosity, and blood pressure then falls because of the decreased viscosity. As you may know, after hemorrhage either whole blood or plasma is preferred to saline solution for transfusions. The reason is that saline solution is not a viscous liquid and so cannot keep blood pressure at a normal level.

No one's blood pressure stays the same all the time. It fluctuates even in a perfectly healthy individual. For example, it goes up when a person exercises strenuously. Not only is this normal but the increased blood pressure serves a good purpose. It tends to increase circulation, to bring more blood to muscles each minute, and thus to supply them with more oxygen and food for more energy.

A normal average arterial blood pressure is 120/80, or 120 millimeters of mercury systolic pressure (as the ventricles contract) and 80 millimeters of mercury diastolic pressure (as the ventricles relax).

Pulse

What you feel when you take a pulse is an artery expanding and then recoiling alternately. To feel a pulse, you must place your fingertips over an artery that lies near the surface of the body and over a bone or other firm background. Six such places are given below. Try to feel your pulse at each of these locations. (You can locate all of them on Fig. 10-6, except the temporal and facial arteries.)

1. *Radial artery*—at the wrist
2. *Temporal artery*—in front of the ear or above and to the outer side of the eye
3. *Common carotid artery*—in the neck

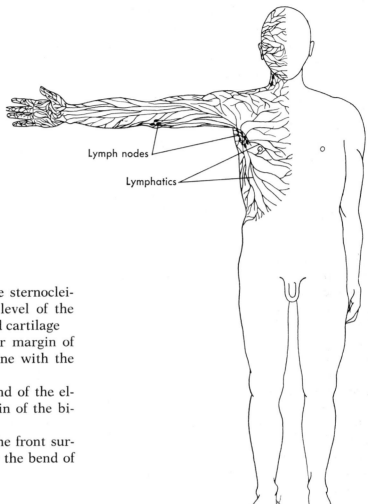

Fig. 10-12 Lymph drainage. The right lymphatic duct drains lymph from the upper right quarter of the body into the right subclavian vein. The thoracic duct drains lymph from all the rest of the body into the left subclavian vein.

Lymph nodes

Lymphatics

along the front edge of the sternocleidomastoid muscle at the level of the lower margin of the thyroid cartilage

4. *Facial artery*—at the lower margin of the lower jawbone on a line with the corners of the mouth

5. *Brachial artery*—at the bend of the elbow along the inner margin of the biceps muscle

6. *Dorsalis pedis artery*—on the front surface of the foot, just below the bend of the ankle joint

Lymphatic system

The lymphatic system is not really a separate system of the body. It is part of the circulatory system, since it consists of lymph, a moving fluid that comes from the blood and returns to the blood by way of the lymphatic vessels. In addition to lymph and the lymphatic vessels, the system includes lymph nodes and specialized lymphatic organs such as the thymus and spleen.

Lymph forms in this way: blood plasma filters out of the capillaries into the microscopic spaces between tissue cells. Here the liquid is called *interstitial fluid* or tissue fluid. Most of the interstitial fluid goes back into the blood by the same route it came out, that is, back through the capillary membrane. The remainder of the interstitial fluid enters tiny lymphatic capillaries to become lymph. It next moves on into larger lymphatics and finally enters the blood in veins in the neck region. The largest lymphatic vessel is the *thoracic duct*. Lymph from about three fourths of the body (Fig. 10-12) eventually drains into the thoracic duct and from it goes back into the blood. The thoracic duct joins the left subclavian vein at the angle where the internal jugular vein also joins it (Fig. 10-7). Lymph from the rest of the body drains into the right lymphatic ducts and on into the right subclavian vein.

Lymph serves a unique transport function by returning tissue fluid, proteins, fats, and other substances to the general circulation. It is important to realize, however, that the flow of lymph differs from the true "circulation" of blood seen in the cardiovascular system. Unlike vessels in the blood vascular system, the lymphatic vessels do not form a closed ring or circuit. Once lymph is formed from interstitial fluid in the soft tissues of the body, it flows only once through its system of lymphatic vessels before draining into the blood through the large neck veins. Lymph does not flow over and over again through vessels that form a circular route.

Lymph nodes

Lymph nodes, or lymph glands as they are sometimes improperly called, are oval structures located mainly in clusters along lymphatics (Fig. 10-12). Some are as small as a pinhead and others as large as a lima bean. The structure of the lymph nodes makes it possible for them to perform two important functions: defense and white blood cell formation.

Defense function: filtration. Fig. 10-13 shows the structure of a typical lymph node. Note that lymph enters the node through four *afferent* (from the Latin "to carry toward") *lymph vessels*. The afferent lymphatics deliver lymph to the node. In passing through the node lymph is filtered so that injurious particles such as bacteria, soot, and cancer cells are removed and prevented from entering the blood and circulating all over the body. Lymph exits from the node through a single *efferent* (from the Latin "to carry away from") *lymph vessel*. Fig. 10-14 shows how bacteria from an infected hair follicle are filtered from lymph passing through nearby nodes.

White blood cell formation. The lymphatic tissue of lymph nodes forms lymphocytes and monocytes, the nongranular white blood cells described earlier in this chapter.

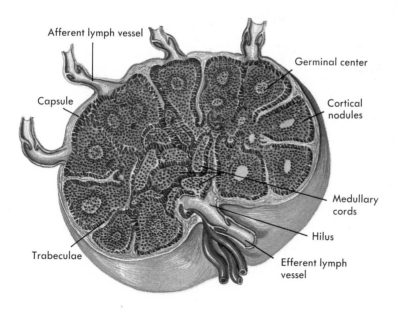

Fig. 10-13 Structure of a lymph node. Several afferent lymphatics bring lymph to the node. Lymph is drained from the node by a single efferent vessel. Lymphatic tissue, densely packed with lymphocytes, composes the substance of the node.

A nurse can use knowledge of lymph node location and function in many ways. For example, when she takes care of a patient with an infected finger, she should watch the elbow and axillary regions for swelling and tenderness of the lymph nodes. These nodes filter lymph returning from the hand and may become infected by the bacteria they trap. A surgeon uses knowledge of lymph node function when removing lymph nodes under the arms (axillary nodes) and in other areas during an operation for breast cancer. These nodes may contain cancer cells filtered out of the lymph drained from the breast. Cancer of the breast is one of the most common forms of

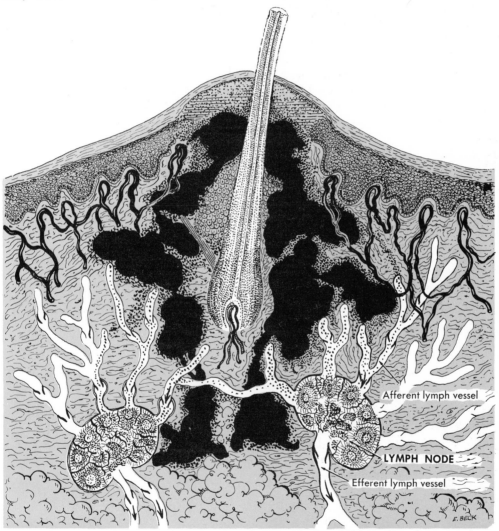

Afferent lymph vessel

LYMPH NODE

Efferent lymph vessel

E. BECK

Fig. 10-14 Diagrammatic representation of a skin section in which an infection surrounds a hair follicle. The black areas and dots around the follicle represent pus and infectious bacteria. Note that lymph entering the node via the afferent lymphatics is contaminated by bacteria. After filtering through the node, lymph in the efferent vessel is bacteria free.

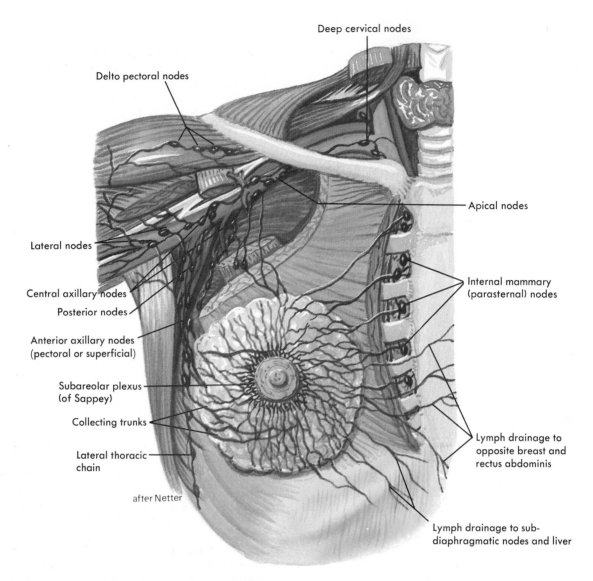

Deep cervical nodes

Delto pectoral nodes

Apical nodes

Lateral nodes

Internal mammary (parasternal) nodes

Central axillary nodes

Posterior nodes

Anterior axillary nodes (pectoral or superficial)

Subareolar plexus (of Sappey)

Collecting trunks

Lymph drainage to opposite breast and rectus abdominis

Lateral thoracic chain

after Netter

Lymph drainage to sub-diaphragmatic nodes and liver

Fig. 10-15 Lymphatic drainage of the breast. Note the extensive network of nodes that receive lymph from the breast.

this disease in women. Unfortunately, cancer cells from a single tumorous growth in the breast often spread to other areas of the body through the lymphatic system. Fig. 10-15 shows how lymph from the breast drains into many different and widely placed nodes.

Spleen

The spleen remains one of the mystery organs of the body. Quite a bit of doubt still exists about its functions. The three lower ribs provide a protective shelter over the spleen, which is located in the upper left corner of the abdominal cavity just under the diaphragm. Except for blood vessels and nerves, the spleen connects with no other organs.

The spleen's main functions seem to be to form lymphocytes and monocytes (as do the lymph nodes) and to act as the body's blood bank. The spleen can store almost 1 pint of blood and quickly release it back into circulation when more blood is needed—during strenuous exercise and after hemorrhage, for instance.

Outline summary

Heart

A Location, size, and position
1 Triangular organ located in mediastinum with two thirds of mass to left of body midline and one third to right; apex on diaphragm; shape and size of a closed fist (see Fig. 10-1)
2 Cardiopulmonary resuscitation (CPR)—heart lies between sternum in front and bodies of thoracic vertebrae behind, rhythmic compression of heart between sternum and vertebrae can maintain blood flow in cardiac arrest; if combined with artificial respiration procedure, can be life-saving
3 Cavities—right atrium, right ventricle, left atrium, left ventricle
4 Wall—myocardium, composed of cardiac muscle
5 Lining—endocardium
6 Covering—pericardium
7 Valves—keep blood flowing in right direction through heart; prevent backflow
  a Tricuspid—at opening of right atrium into ventricle
  b Mitral (or bicuspid—at opening of left atrium into ventricle
  c Pulmonary semilunars—at beginning of pulmonary artery
  d Aortic semilunars—at beginning of aorta
8 Conduction system—see Fig. 10-3
  a SA (sinoatrial) node, the pacemaker—located in the wall of the right atrium near the opening of the superior vena cava
  b AV (atrioventricular) node—located in the right atrium along the lower part of the interatrial septum
  c AV (bundle of His) bundle—located in the septum of the heart
  d Purkinje fibers—located in the walls of the ventricles

Blood vessels

A Kinds
1 Arteries—carry blood away from heart
2 Veins—carry blood toward heart
3 Capillaries—carry blood from arterioles to venules
B Structure—see Figs. 10-4 and 10-5
C Functions
D Names of main arteries—see Fig. 10-6
E Names of main veins—see Fig. 10-7

Circulation

A Plan of circulation—see Fig. 10-9
B Portal circulation—detour of venous blood from stomach, pancreas, intestines, spleen through liver before return to heart
C Fetal circulation—see Fig. 10-10

Blood pressure

A Blood pressure is push, or force of blood in blood vessels
B Highest in arteries, lowest in veins—see Fig. 10-11
C Blood pressure gradient causes blood to circulate—liquids can flow only from area where pressure is higher to where it is lower
D Blood volume, heartbeat, and blood viscosity are main factors that produce blood pressure
E Blood presssure varies within normal range from time to time

Pulse

A Definition—alternate expansion and recoil of blood vessel wall
B Places where you can count the pulse easily—see p. 156

Lymphatic system (see Figs. 10-12 to 10-15)

A Consists of lymphatic vessels, lymph nodes, lymph, and spleen
B Lymph—the fluid in the lymphatic vessels; lymph comes from blood by plasma filtering

fluid, some of which then enters the lymph capillaries to become lymph and be returned to blood by way of lymphatics; largest lymphatic is thoracic duct—drains lymph from all but upper right quarter of body into left subclavian vein
C Lymph nodes
  1 Located along certain lymphatics, usually in clusters; for example, at elbow, under arm, in groin, at knee
  2 Functions—filter out injurious particles such as microorganisms and cancer cells from lymph before it returns to blood; form some white blood cells (lymphocytes and monocytes)

Spleen

A Forms some white blood cells (lymphocytes and monocytes)
B Serves as blood bank for body—stores blood until needed and then releases it back into circulation

New words

afferent
artificial pacemaker
cardiac output
cardiopulmonary
  resuscitation (CPR)
diastolic pressure
efferent
embolism

endocarditis
myocardial infarction
myocarditis
pericarditis
portal circulation
systemic pressure
thrombosis

Review questions

1 Describe the position of the heart in the mediastinum.
2 What is cardiopulmonary resuscitation (CPR)?
3 Briefly explain what happens when blood clots, including what makes it start to clot.
4 You hear that a friend has a "coronary thrombosis." What does this mean to you?
5 Describe blood flow through the heart.
6 A patient has had an operation to repair the mitral valve. Where is this valve, and what is its function?
7 What are some differences between an artery, a vein, and a capillary?
8 Considering that the function of the circulatory system is to transport substances to and from the cells, do you think it is true that in one sense, capillaries are our most important blood vessels? Give a reason for your answer.
9 The right ventricle of the heart pumps blood to and through only one organ. Which one?
10 What part of the heart pumps blood through the systemic circulation, that is, to and through all organs other than the lungs?
11 All blood returns from the systemic circulation to what part of the heart?
12 What part of the heart pumps blood through the pulmonary circulation, that is, to and through the lungs?

13 Blood returning from the pulmonary circulation (from the lungs, in other words) enters what part of the heart?

14 From which cavity of the heart does blood rich in oxygen leave the heart to be delivered to tissue capillaries all over the body?

15 How do arterial blood and venous blood differ with regard to their oxygen and carbon dioxide contents?

16 Does every artery carry arterial blood and every vein carry venous blood? If not, what exceptions are there?

17 Explain what is meant by "portal circulation."

18 Name the vein at the bend of the elbow into which substances are often injected and from which blood is sometimes withdrawn. (Check your answer with Fig. 10-8.)

19 Nurses frequently have to take a patient's blood pressure. What vital function does blood pressure perform?

20 Sometimes a woman's arm becomes very swollen for a while after removal of a breast and the nearby lymph nodes and lymphatics, including some of those in the upper arm. Can you think of any reason why swelling occurs?

21 You could live without your spleen since it does not do anything vital for the body. What functions does it perform?

22 Discuss the two functions of lymph nodes.

23 How does the "circulation" of lymph differ from blood circulation?

chapter 11

The immune system

The immune system is the body's defense system. It defends us against three major kinds of enemies: microorganisms that invade our bodies, foreign tissue cells that have been transplanted into our bodies, and our own cells that have turned malignant (cancerous). The immune system functions to make us immune, that is, able to resist these enemies. Unlike other systems of the body, which are made up of groups of organs, the immune system is made up of billions of cells and trillions of molecules. This chapter presents information about these cells and molecules.

Immune system cells

The most numerous cells of the immune system are white blood cells of the type called lymphocytes. A million million strong, lymphocytes continually patrol the body, searching out any enemy cells that may have entered. Lymphocytes circulate in the body's fluids. Huge numbers of them wander vigilantly through most of its tissues. Lymphocytes densely populate the body's widely scattered lymph nodes and its other lymphatic tissues—especially the thymus gland in the chest and the spleen and liver in the abdomen.

In addition to the trillion or so lymphocytes, another large group of cells known collectively as phagocytes also functions as part of the immune system. Phagocytes are cells that can carry on phagocytosis (ingestion and digestion) of foreign cells or particles. They include two types of white blood cells, neutrophils and monocytes, and connective tissue cells named macrophages.

Immune system molecules

The immune system consists not only of cells but also of certain protein molecules. Chief among them are the proteins called antibodies and complement. The number of immune system molecules far exceeds the number of immune system cells. Antibody molecules alone outnumber the lymphocytes by about 100 million to one. How antibodies and complement function to defend us against invaders is described on pp. 168 and 169.

Lymphocytes

Types

There are two major types of lymphocytes, designated as B lymphocytes and T lymphocytes but usually called B cells and T cells.

Development of B cells

All lymphocytes that circulate in the tissues arise from primitive cells in the bone marrow called stem cells and go through two stages of development. In chickens the first stage of B cell development occurs in a structure named the bursa of Fabricus—

hence the name "B cells." Humans have no such bursa, and the structure in which B cells undergo their first stage of development is appropriately called the "bursa-equivalent structure." Its identity has not yet been positively established. Persuasive evidence, however, indicates the liver. The first stage of B cell development consists of stem cells becoming transformed into immature B cells. The process begins shortly before birth and is completed by the time a human infant is a few months old. *Immature B cells* are small lymphocytes that have synthesized and inserted into their cytoplasmic membranes numerous molecules of one specific kind of antibody (see Fig. 11-1). These antibody-bearing immature B cells leave the bursa-equivalent structure where they were formed, enter the blood, and are transported to their new place of residence, chiefly the lymph nodes. There they act as seed cells. Each immature B cell undergoes repeated mitosis (cell division) and forms a clone of immature B cells. A *clone*, by definition, is a family of many identical cells all descended from one cell. Because all the cells in a clone of immature B cells have descended from one immature B cell, all of them bear the same surface antibody molecules as did their single ancestor cell.

The second stage of B cell development changes an immature B cell into an activated B cell. Not all immature B cells undergo this change. They do so only if an immature B cell comes in contact with certain protein molecules—called antigens—whose

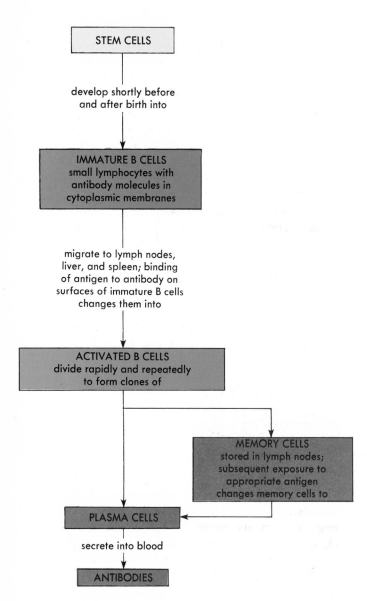

Fig. 11-1 B cell development takes place in two stages. First stage: Shortly before and after birth, stem cells develop into immature B cells. This occurs in bursa-equivalent structure, perhaps the liver. Second stage (occurs only if immature B cell contacts its specific antigen): Immature B cell develops into activated B cell, which divides rapidly and repeatedly to form a clone of plasma cells and a clone of memory cells. Plasma cells secrete antibodies capable of combining with specific antigen that caused immature B cell to develop into active B cell.

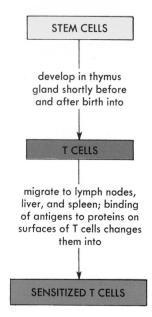

Fig. 11-2 T cell development. First stage takes place in thymus gland shortly before and after birth. Second stage occurs only if T cell contacts antigen, which combines with certain proteins present on T cell's surface.

shape fits the shape of the immature B cell's surface antibody molecules. If this happens, the antigens lock onto the antibodies and by so doing change the immature B cell into an activated B cell. Then the activated B cell, by dividing rapidly and repeatedly, develops into clones of two kinds of cells, namely, plasma cells and memory cells. *Plasma cells* secrete copious amounts of antibody into the blood—reportedly, 2,000 antibody molecules per second by each plasma cell for every second of the few days that it lives. Antibodies circulating in the blood constitute an enormous, mobile, ever-on-duty army. How antibodies function is described on p. 168.

Memory cells have the ability to secrete antibodies but do not immediately do so. They remain in reserve in the lymph nodes until such time as they may be contracted by the same antigen that led to their formation. Then, very quickly, the memory cells develop into plasma cells and secrete large amounts of antibody. Memory cells, in effect, seem to remember their ancestor-activated B cell's encounter with its appropriate antigen. They stand ready, at a moment's notice, to produce antibody that will combine with this antigen.

Development of T cells

T cells, by definition, are lymphocytes that have undergone their first stage of development in the thymus gland. Stem cells from the bone marrow seed the thymus, and shortly before and after birth they develop into T cells. The newly formed T cells stream out of the thymus into the blood and migrate chiefly to the lymph nodes where they take up residence. Embedded in each T cell's cytoplasmic membrane are protein molecules so shaped that they can fit only one specific kind of antigen molecule. The second stage of T cell development takes place when and if a T cell comes into contact with its specific antigen. If this happens, the antigen binds to the protein on the T cell's surface, thereby changing the T cell into a sensitized T cell (see Fig. 11-2).

Function of B cells

B cells function indirectly to produce humoral immunity. Humoral immunity is resistance to disease organisms produced by the actions of antibodies circulating in body fluids. Activated B cells develop into plasma cells. Plasma cells secrete antibodies into the blood; they are the "antibody factories" of the body. How antibodies act to produce immunity is described on p. 168.

Functions of T cells

Sensitized T cells function to produce cell-mediated immunity. As the name sug-

gests, *cell-mediated immunity* is resistance to disease organisms resulting from the actions of cells—chiefly sensitized T cells. Some sensitized T cells kill invading cells directly. When bound to antigens present on an invading cell's surface, they release a substance called *lymphotoxin*, which acts as a specific and lethal poison against the bound cell. Many sensitized cells produce their deadly effects indirectly by means of compounds known collectively as lymphokines, which they release into the area around enemy cells. Among these are a substance called chemotactic factor and another called macrophage activating factor. *Chemotactic factor* attracts macrophages into the neighborhood of the enemy cells. Then *macrophage activating factor* prods the assembled macrophages into destroying the cells by phagocytosing (ingesting and digesting) them.

Antibodies (immunoglobulins)
Definitions

Antibodies are protein compounds normally present in the body. A defining characteristic of an antibody molecule is the presence on its surface of two uniquely shaped concave regions called *combining sites*. Another defining characteristic is an antibody molecule's ability to combine with a specific compound called an antigen. An *antigen* is a compound whose molecules have small regions on their surfaces—called *epitopes*—uniquely shaped so as to fit into the combining sites of a specific antibody molecule as precisely as a key fits into a specific lock. Antigens are usually foreign proteins. Most often antigens are molecules present in the surface membranes of invading cells, such as microorganisms, cancer cells, or transplanted tissue cells.

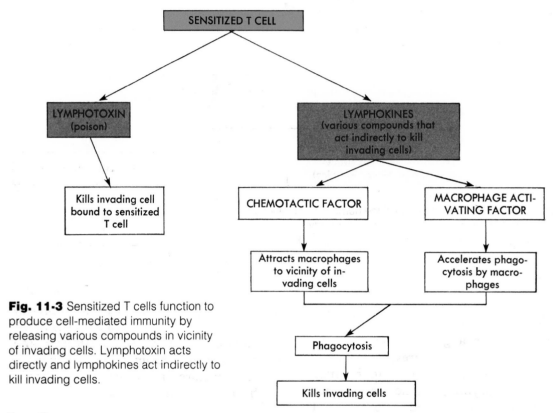

Fig. 11-3 Sensitized T cells function to produce cell-mediated immunity by releasing various compounds in vicinity of invading cells. Lymphotoxin acts directly and lymphokines act indirectly to kill invading cells.

Functions

In general, antibodies function to produce immunity by making antigens unable to harm the body. To do this an antibody must first bind to its specific antigen, the one whose epitopes fit its combining sites. This forms an antigen-antibody complex. The antigen-antibody complex then acts in one or more ways to make the antigen, or the cell on which it is present, harmless. For example, if the antigen is a toxin, a substance poisonous to body cells, the toxin is neutralized, made nonpoisonous, by becoming part of an antigen-antibody complex. Or if antigens are molecules in the surface membranes of invading cells, when antibodies combine with them, the resulting antigen-antibody complexes may agglu-tinate the enemy cells, that is, make them stick together in clumps. Then macrophages or other phagocytes can rapidly destroy them by ingesting and digesting large numbers of them at one time. But probably the most important way in which antibodies act is one we will consider last. In many instances, when antigens that are molecules on a cell's surface combine with antibody molecules, they change the shape of the antibody molecule just slightly, but enough to expose two previously hidden regions. These are called complement-binding sites. Their exposure initiates a series of events that kill the cell on whose surface they take place. The next section describes these events.

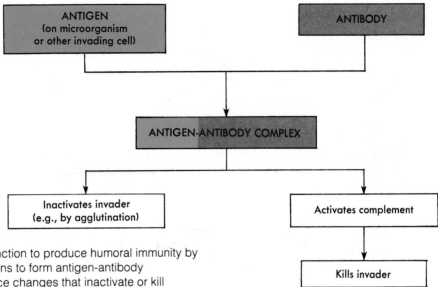

Fig. 11-4 Antibodies function to produce humoral immunity by binding to specific antigens to form antigen-antibody complexes, which produce changes that inactivate or kill invading cells.

Complement

Definition

Complement is a group of inactive enzyme proteins normally present in blood. Formerly there were thought to be nine complement enzymes, but newer evidence indicates there may be eleven or more of them. They are identified by numbers that indicate their order of functioning.

Activation

Very briefly, complement is activated by antibodies. It begins with the binding of antibodies to antigens located on an invading cell's surface. This changes the shape of the antibody molecule and thereby exposes its two complement-binding sites. This in turn activates complement enzyme 1 to bind to the exposed complement binding sites, an event that activates complement enzyme 2 to bind to an adjacent antibody molecule. Then in rapid-fire succession the remaining complement enzymes are similarly activated and bound to antibody molecules. Together they form a doughnut-shaped assemblage—complete with a hole in the middle! Keep in mind that the antibody molecules are bound both to antigens in the invading cell's cytoplasmic membrane and to complement enzymes. The doughnut-shaped structure thus formed is fixed firmly on the invading cell's surface.

Function

Complement functions to kill invading cells of various types. How? By, in effect, drilling a hole in their cytoplasmic membranes! Fluid then pours through the hole into the cell, swelling it, and soon bursting it wide open.

Outline summary
Immune system cells

A Lymphocytes—type of white blood cells; most numerous of immune cells, estimated 1 trillion lymphocytes in body.

B Phagocytes—cells that carry on phagocytosis, that is, ingestion and digestion of foreign cells or particles; include two types of white blood cells, neutrophils and monocytes, and connective tissue cells called macrophages.

Immune system molecules

Protein molecules, chiefly antibodies and complement; estimated 100 million times as many antibody molecules as lymphocytes

Lymphocytes

A Types—B cells and T cells
B Development of B cells—primitive stem cells migrate from bone marrow to "bursa-equivalent" structure, perhaps the liver
  1 First stage—stem cells develop into immature B cells; takes place in bursa-equivalent structure during the few months before and after birth; immature B cells are small lymphocytes with antibody molecules (which they have synthesized) in their cytoplasmic membranes; migrate chiefly to lymph nodes
  2 Second stage—immature B cell develops into activated B cell; initiated by immature B cell's contact with antigens, which bind to its surface antibodies; activated B cell, by dividing repeatedly, forms two clones of cells: plasma cells and memory cells; plasma cells secrete antibodies into blood; memory cells stored in lymph nodes; if subsequent exposure to antigen that activated B cell occurs, memory cells become plasma cells and secrete antibodies
C Development of T cells—stem cells from bone marrow migrate to thymus gland
  1 Stage 1—stem cells develop into T cells; occurs in thymus during few months before and after birth; T cells migrate chiefly to lymph nodes
  2 Stage 2—T cells develop into sensitized T cells; occurs when, and if, antigen binds to T cell's surface proteins
D Function of B cells—indirectly, B cells produce humoral immunity; activated B cells develop into plasma cells; plasma cells secrete antibodies into blood; circulating antibodies produce humoral immunity.
E Functions of T cells—produce cell-mediated immunity; kill invading cells by releasing lymphotoxin that poisons cells and also by releasing chemicals that attract and activate macrophages to kill cells by phagocytosis

Antibodies (immunoglobulins)

Antibodies bind to antigens to form antigen-antibody complexes that act in several ways to make antigens and cells with surface antigens harmless; for example, neutralization of toxins, agglutination of invading cells; binding of antigen to antibody changes shape of antibody molecule, exposing its complement-binding sites

Complement

A Definition—group of inactive enzymes normally present in blood
B Activation—complement activated by binding to antibodies that are already bound to antigens on surface of invading cell; activated complement enzymes assemble in form of doughnut-shaped structure on cell surface; hole in center of this structure
C Function—complement kills invading cells by "drilling" hole in their cytoplasmic membranes, which allows fluid to enter cell until it bursts

Review questions

1 Explain what the terms "antigens" and "antibodies" mean.
2 What are combining sites? complement binding sites? epitopes?
3 Name four kinds of cells that constitute part of the immune system. Which are most numerous? How numerous are they estimated to be?
4 Name the two major immune system molecules.
5 How does the number of antibodies compare with the number of lymphocytes?
6 What are the two major types of lymphocytes called? What makes these names appropriate?
7 Describe briefly the two stages of development that B cells undergo.
8 Describe briefly the two stages of development that T cells undergo.
9 What cells secrete antibodies?
10 Explain the function of memory cells.
11 Describe the function of activated B cells.
12 Describe the function of sensitized T cells.
13 Immunoglobulins is a synonym for what word?

Systems that process and distribute foods and eliminate wastes

chapter 12

The digestive system

The digestive system's main organs form an irregular-shaped tube, open at both ends, called the alimentary canal or the gastrointestinal (GI) tract. Located in the tract or opening into it are several digestive organs. Table 12-1 names both main and accessory digestive organs.

Foods undergo three kinds of processing in the body: digestion, absorption, and metabolism. Digestion and absorption are functions performed by the organs of the digestive system. Metabolism, on the other hand, is a function performed by all body cells. In this chapter we shall begin by describing digestive system organs and then discuss digestion, absorption, and metabolism.

Mouth

A roof, floor, and side walls form the mouth (oral or buccal cavity). The roof consists of the hard and soft palates. The hard palate is hard because it is composed of bone, namely, the palatine bones and parts of the maxillary bones. The soft palate is soft because it consists chiefly of muscle; it is an arch-shaped structure that serves as a curtain separating the mouth from the nasopharynx (the part of the throat behind the nose). The archway of the soft palate opens into the oropharynx (the part of the throat behind the mouth). Hanging down from the center of the soft palate is a cone-shaped process named the uvula. If you look in the

Table 12-1 Organs of the digestive system

Main organs	Accessory organs
Mouth	Teeth
	Salivary glands (parotid, sublingual, submandibular)
Pharynx (throat)	
Esophagus (food pipe)	
Stomach	
Small intestine Duodenum	Liver and gallbladder
	Pancreas
Jejunum	
Ileum	
Large intestine Cecum	Appendix
Colon (ascending, transverse, descending, and sigmoid)	
Rectum	

mirror, open your mouth wide, and say "Ah," you can see your uvula.

The floor of the mouth consists of the tongue and its muscles. The tongue is made of skeletal muscle covered with mucous membrane. It is attached to four bones, namely, the mandible, two temporal bones, and the hyoid. Have you ever noticed the many small elevations on the surface of your tongue? They are named papillae; nerve endings called taste buds are located in many of the papillae. A fold of mucous membrane, the frenulum, helps anchor the tongue to the floor of the mouth. Occasionally the frenulum is too short to allow free movements of the tongue. Such individuals cannot enunciate words normally and are said to be tongue-tied.

Fig. 12-1 The deciduous arch. Note that in the set of twenty temporary primary teeth there are no premolars (bicuspids) and there are only two pairs of molars in each jaw. Compare with permanent teeth shown in Fig. 12-2.

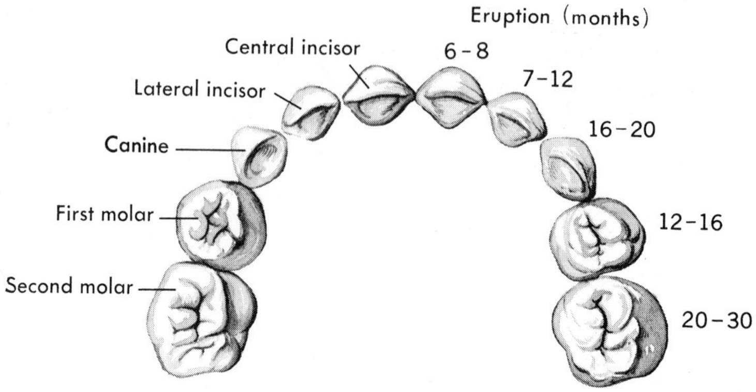

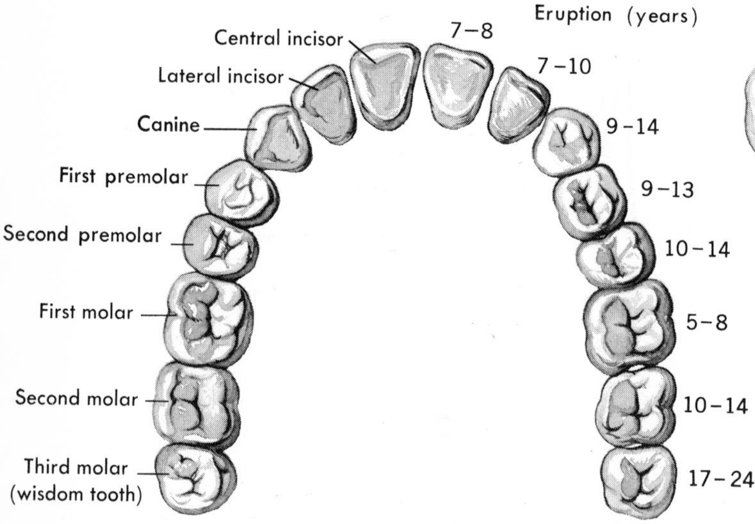

Eruption (months)

Central incisor — 6-8
Lateral incisor — 7-12
Canine — 16-20
First molar — 12-16
Second molar — 20-30

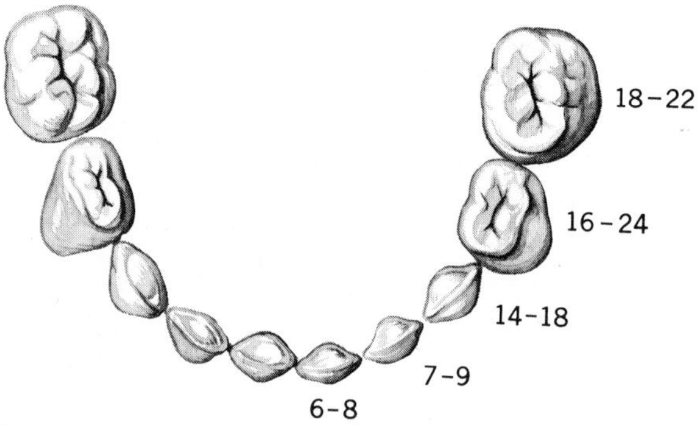

Eruption (years)

Central incisor — 7-8
Lateral incisor — 7-10
Canine — 9-14
First premolar — 9-13
Second premolar — 10-14
First molar — 5-8
Second molar — 10-14
Third molar (wisdom tooth) — 17-24

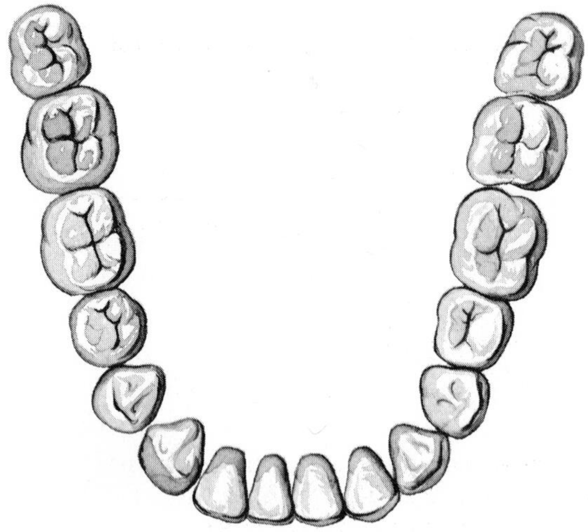

18-22
16-24
14-18
7-9
6-8

Fig. 12-2 The thirty-two permanent teeth. Generally the lower teeth erupt before the corresponding upper teeth, and all teeth usually erupt earlier in girls than in boys.

Teeth

By the time a baby is 2 years old, he probably has his full set of twenty baby teeth. When a young adult is somewhere between 17 and 24 years old, he usually has his full set of thirty-two permanent teeth. The average age for cutting the first tooth is about 6 months, and the average age for losing the first baby tooth and starting to cut the permanent teeth is about 6 years. Figs. 12-1 and 12-2 give the names of the teeth and show which ones are lacking in the deciduous or baby set. Fig. 12-3 illustrates tooth structure.

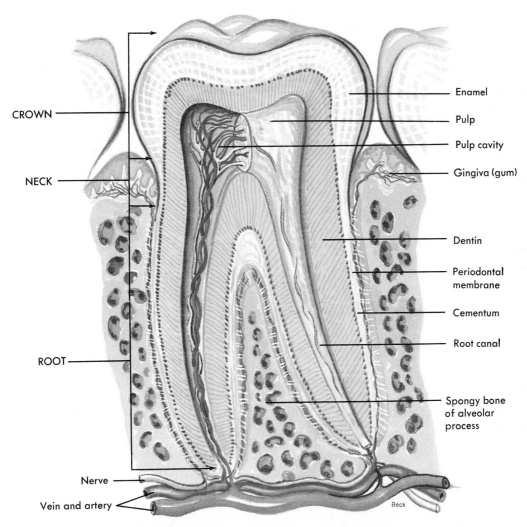

CROWN

NECK

ROOT

Enamel

Pulp

Pulp cavity

Gingiva (gum)

Dentin

Periodontal membrane

Cementum

Root canal

Spongy bone of alveolar process

Nerve

Vein and artery

Beck

Fig. 12-3 A molar tooth sectioned to show its bony socket and details of its three main parts: crown, neck, and root. Enamel (over the crown) and cementum (over the neck and root) surround the dentin layer. The pulp contains nerves and blood vessels.

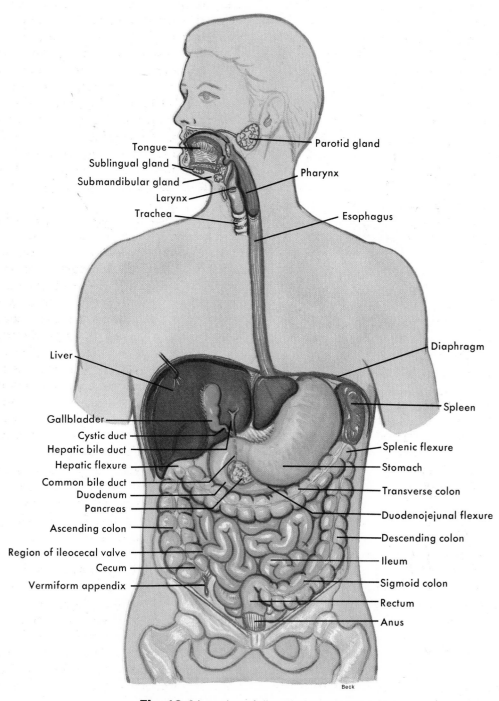

Tongue
Sublingual gland
Submandibular gland
Larynx
Trachea
Parotid gland
Pharynx
Esophagus

Liver
Diaphragm
Spleen

Gallbladder
Cystic duct
Hepatic bile duct
Hepatic flexure
Common bile duct
Duodenum
Pancreas
Ascending colon
Region of ileocecal valve
Cecum
Vermiform appendix

Splenic flexure
Stomach
Transverse colon
Duodenojejunal flexure
Descending colon
Ileum
Sigmoid colon
Rectum
Anus

Beck

Fig. 12-4 Location of digestive system organs.

Salivary glands

The parotid glands lie just below and in front of each ear at the angle of the jaw—an interesting anatomical fact because it explains why people who have mumps (an infection of the parotid gland) often complain that it hurts when they open their mouths or chew; these movements squeeze the tender inflamed gland. To see the openings of the ducts of the parotid glands, look on the insides of your cheeks opposite the second molar tooth, on either side of the upper jaw. Besides the parotid glands there are also two other pairs of salivary glands—the submandibular and the sublingual glands (Fig. 12-4). Their ducts open into the floor of the mouth.

Pharynx

The pharynx is a tubelike structure made of muscle and lined with mucous membrane. Observe its location in Fig. 12-4. Because of its location behind the nasal cavities and mouth, it functions as part of both the respiratory and digestive systems. Air must pass through the pharynx on its way to the lungs and food must pass through it on its way to the stomach.

Esophagus

The esophagus or food pipe is the muscular, mucous-lined tube that connects the pharynx with the stomach. It is about 25 centimeters (10 inches) long.

Stomach

The stomach lies in the upper part of the abdominal cavity just under the diaphragm. It serves as a pouch that food enters after it has been chewed, swallowed, and passed through the esophagus. The stomach looks small after it is emptied, not much bigger than a large sausage, but it expands considerably after a large meal. Have you ever felt so uncomfortably full after eating that you could not take a deep breath? If so, it probably meant that your stomach was so full of food that it occupied more space than usual and pushed up against the diaphragm. This made it hard for the diaphragm to contract and move downward as much as necessary for a deep breath.

Three layers of smooth muscle, with fibers running lengthwise, around, and obliquely in the stomach wall, make the stomach one of the strongest internal organs—well able to break up food into tiny particles and to mix them thoroughly with the gastric juice. Stomach muscle contractions take part in producing *peristalsis*, the movement that propels food down the length of the digestive tract. Mucous membrane lines the stomach; it contains thousands of microscopic glands that secrete gastric juice and hydrochloric acid into the stomach. When the stomach is empty, its linings lies in folds called *rugae*.

The lower part of the stomach is called the *pylorus*. It is a narrow section that joins the first part of the small intestine (the *duodenum*). Food is held in the stomach by the pyloric sphincter muscle long enough for partial digestion. The sphincter consists of circular smooth muscle fibers that stay contracted most of the time and thereby close off the opening of the pylorus into the duodenum. The fibers relax at intervals when part of the food is ready to leave the stomach, but sometimes they go into a spasm and do not relax normally; this condition is referred to as *pylorospasm*. It occurs fairly often in babies.

Small intestine

The small intestine seems to be mis-named if you look only at its length—it is roughly 20 feet long. It is noticeably smaller around, however, than the large intestine; so in this respect its name is appropriate. Different names identify different sections of the small intestine. In the order in which food passes through them, they are the *duodenum, jejunum,* and *ileum.*

The mucous lining of the small intestine, like that of the stomach, contains thousands of microscopic glands. They are called *intestinal glands,* and they secrete the intestinal digestive juice. Another structural feature of the lining of the small intestine makes it especially well suited to its function of food and water absorption; it is not perfectly smooth as it appears to the naked eye. Instead it has thousands of tiny "fingers," called *villi.* Under the microscope

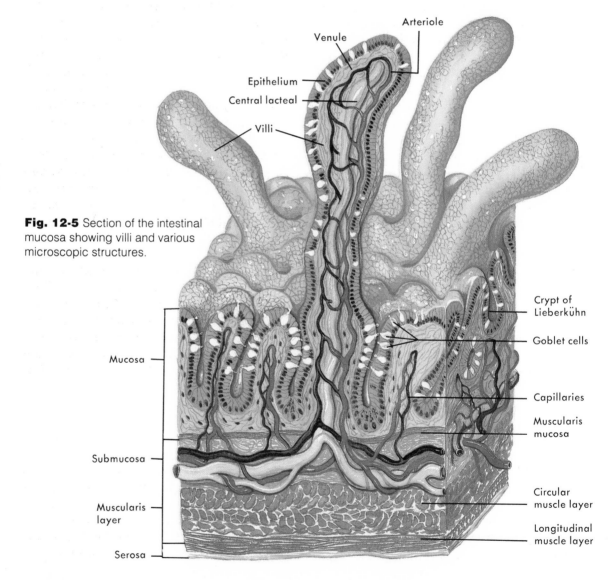

Fig. 12-5 Section of the intestinal mucosa showing villi and various microscopic structures.

they can be seen projecting into the hollow interior of the intestine. Inside each villus lies a rich network of blood and lymph capillaries. See Fig. 12-5. Millions and millions of villi jut inward from the mucous lining. Imagine the lining as perfectly smooth without any villi; think how much less surface area there would be for contact between capillaries and intestinal lining. Consider what an advantage a large contact area offers for faster absorption of food from the intestine into the blood and lymph—one more illustration of the principle that structure determines function.

Smooth muscle in the wall of the small intestine contracts to produce peristalsis, the wormlike movements that move food through the tract.

Liver and gallbladder

The liver is such a large organ that it fills the entire upper right section of the abdominal cavity and even extends partway over onto the left side. Because its cells secrete a substance (bile) into ducts, the liver is classified as an exocrine gland—in fact, it is the largest gland in the body.

Look now at Fig. 12-6. First, identify the *hepatic ducts.* They drain bile out of the liver, a fact suggested by the name hepatic, which comes from the Greek word for liver *(hepar).* Next, notice the duct that drains bile into the small intestine (duodenum), namely the *common bile duct.* It is formed by the union of the *common hepatic duct* with the *cystic duct.* Between meals, much of the bile moves up the cystic duct into the gallbladder (located on the undersurface of the liver) for concentration and storage. After meals, when fats enter the duodenum, the gallbladder ejects bile into the cystic duct from which it drains down the common bile duct and on into the duo-

denum. Visualize a gallstone blocking the common bile duct in Fig. 12-6. Bile could not then drain into the duodenum and leave the body in the feces. Therefore excessive amounts of bile would be absorbed into the blood. A yellowish skin discoloration called jaundice would result. Obstruction of the common hepatic duct also leads to jaundice. Because bile cannot then drain out of the liver, excessive amounts of it are absorbed. Do you think obstruction of the cystic duct would lead to jaundice? Why not?

In addition to secreting about a pint of bile a day, liver cells perform other functions necessary for healthy survival. They play a major role in the metabolism of all three kinds of foods. They help maintain a normal blood glucose concentration by carrying on complex and essential chemical reactions—glycogenesis, for example. (See p. 185.) Liver cells also carry on the first steps of protein and fat metabolism and synthesize several kinds of protein compounds. They release them into the blood where they are called the blood proteins or plasma proteins. Prothrombin and fibrinogen, two of the plasma proteins formed by liver cells, play essential parts in blood clotting. (See p. 136.) Another protein made by liver cells, albumin, helps maintain normal blood volume. Liver cells detoxify various poisonous substances such as bacterial products and certain drugs. Liver cells store several substances, notably, iron and vitamins A, B_{12}, and D.

Pancreas

The pancreas lies behind the stomach, cradled "in the arms of the duodenum." It is both an exocrine gland that secretes the pancreatic juice into ducts and an endocrine gland that secretes hormones into the blood. Pancreatic juice is the most impor-

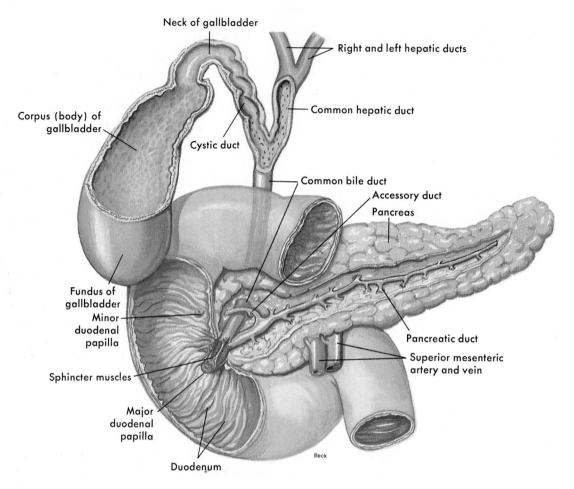

Fig. 12-6 The gallbladder and bile ducts. Obstruction of either the hepatic or common bile duct by stone or spasm blocks the exit of the bile from the liver, where it is formed, and prevents bile from being ejected into the duodenum.

tant digestive juice. It contains enzymes that digest all three major kinds of foods. It also contains sodium bicarbonate, an alkaline substance that neutralizes the hydrochloric acid in the gastric juice that enters the intestines. Pancreatic juice enters the duodenum of the small intestine at the same place that bile enters. As you can see in Fig. 12-6, both the common bile duct and the pancreatic duct open into the duodenum at the major duodenal papilla.

Between the cells that secrete pancreatic juice into ducts lie clusters of cells that have no contact with any ducts. These are the *islands of Langerhans*. They consist of two different kinds of cells—beta cells, which secrete insulin, and alpha cells, which secrete glucagon. (See p. 124 for the function of these hormones.)

Large intestine

The large intestine forms the lower part of the digestive tract. Its divisions—cecum, ascending colon, transverse colon, descending colon, sigmoid colon, and rectum—are shown in Fig. 12-4. The rectum is about 7 or 8 inches long. Its external opening is called the *anus*. Two sphincter muscles stay contracted to keep the anus closed except during defecation. Smooth, or involuntary, muscle composes the inner anal sphincter, but striated, or voluntary, muscle composes the outer one. This anatomical fact sometimes becomes highly important from a practical standpoint. For example, often after a person has had a stroke, the voluntary anal sphincter at first becomes paralyzed. This means, of course, that the individual has no control at this time over bowel movements. Or, in hospital language, the patient has "involuntary defecations."

Appendix

The appendix is a dead-end tube off the cecum. It serves no known function in humans. It is, however, often a nuisance when its mucous lining becomes inflamed—the well-known affliction, appendicitis.

Peritoneum

The *peritoneum* is a large, moist, slippery sheet of serous membrane that lines the abdominal cavity and covers the organs located in it, including most of the digestive organs. The *parietal layer* of the peritoneum lines the abdominal cavity. The *visceral layer* of the peritoneum forms the outer or covering layer of each abdominal organ. The small space between the parietal and visceral layers is called the *peritoneal space*. It contains just enough peritoneal fluid to keep both layers of the peritoneum moist and able to slide freely against each other during breathing and digestive movements.

Extensions

The two most prominent extensions of the peritoneum are named the mesentery and the greater omentum. The *mesentery*, an extension of the parietal peritoneum, is shaped like a giant, plaited fan. Its smaller edge attaches to the lumbar region of the posterior abdominal wall and its long, loose outer edge encloses most of the small intestine, anchoring it to the posterior abdominal wall. The *greater omentum* is an extension of the visceral peritoneum from the lower edge of the stomach, part of the duodenum, and transverse colon. Shaped like a large apron, it hangs down over the intestines, and because spotty deposits of fat give it a lacy appearance, the greater omentum has been nicknamed the "lace apron." It usually envelops a badly inflamed appendix, walling it off from the rest of the abdominal organs.

Digestion

Digestion, a complex process that occurs in the alimentary canal, consists of various physical and chemical changes that prepare food for absorption. *Mechanical digestion* breaks food into tiny particles, mixes them with digestive juices, moves them along the alimentary canal, and finally eliminates the digestive wastes from the body. Chewing (mastication), swallowing (deglutition), peristalsis, and defecation are the main processes of mechanical digestion. *Chemical digestion* breaks down large, nonabsorbable food molecules into smaller, absorbable molecules—molecules that are able to pass through the intestinal mucosa into blood and lymph. Chemical digestion consists of numerous chemical reactions catalyzed by enzymes present in the following digestive juices: saliva, gastric juice, pancreatic juice, and intestinal juice.

Carbohydrate digestion

Very little digestion of carbohydrates (starches and sugars) occurs before food reaches the small intestine. The enzymes in saliva usually have too little time to do their work because so many of us swallow our food so fast. Gastric juice contains no carbohydrate-digesting enzymes. But once the food reaches the small intestine, pancreatic and intestinal juice enzymes digest the starches and sugars. A pancreatic enzyme (amylase) starts the process by changing starches into a sugar, namely, maltose. Three intestinal enzymes—maltase, sucrase, and lactase—digest sugars by changing them into simple sugars, chiefly glucose (dextrose). Maltase digests maltose (malt sugar), sucrase digests sucrose (ordinary cane sugar), and lactase digests lactose (milk sugar). The end products of carbohydrate digestion are the so-called simple sugars, the most abundant of which is glucose.

Protein digestion

Protein digestion starts in the stomach. Two enzymes (rennin and pepsin) in the gastric juice cause the giant protein molecules to break up into somewhat simpler compounds. Then in the intestine other enzymes (trypsin in the pancreatic juice and peptidases in the intestinal juice) finish the job of protein digestion. Every protein molecule is made up of many amino acids joined together. When enzymes have split up the large protein molecule into its separate amino acids, protein digestion is completed. Hence the end product of protein digestion is amino acids. For obvious reasons, amino acids are also referred to as "protein building blocks."

Fat digestion

Not only very little carbohydrate digestion but also very little fat digestion occurs before food reaches the small intestine. An enzyme in gastric juice (gastric lipase) digests some fat in the stomach, but most fats go undigested until after bile in the duodenum emulsifies them—that is, breaks the fat droplets into very small droplets. After this takes place, the pancreatic enzyme (steapsin or pancreatic lipase) splits up the fat molecules into fatty acids and glycerol (glycerin). The end products of fat digestion, then, are fatty acids and glycerol.

Table 12-2 summarizes the main facts about chemical digestion. Enzyme names indicate the type of food digested by the enzyme. For example, the name *amylase* indicates that the enzyme digests carbohydrates (starches and sugars). The name *protease* indicates a protein-digesting enzyme, and the name *lipase* means a fat-digesting enzyme. When carbohydrate digestion has been completed, starches and complex sugars (polysaccharides) have been changed mainly to glucose, a simple sugar (monosaccharide). The end product of protein digestion, on the other hand, is amino acids. Fatty acid and glycerol are the end products of fat digestion. Use information in Table 12-2 to answer questions 10 to 15, p. 189.

Absorption

After food is digested, it is absorbed; that is, it moves through the mucous membrane lining of the small intestine into the blood and lymph. In other words, food absorption is the process by which molecules of amino acids, glucose, fatty acids, and glycerol go from the inside of the intestines into the circulating fluids of the body. Absorption of foods is just as essential a process as digestion of foods. The reason is fairly obvious. As long as food stays in the intestines, it cannot nourish the millions of cells composing all other parts of the body. Their lives

Table 12-2 Chemical digestion

Digestive juices and enzymes	Enzyme digests (or hydrolyzes)	Resulting product*
Saliva Amylase (ptyalin)	Starch (polysaccharide or complex sugar)	Maltose (a disaccharide or double sugar)
Gastric juice Protease (pepsin) plus hydrochloric acid	Proteins, including casein	Proteoses and peptones (partially digested proteins)
Lipase (of little importance)	Emulsified fats (butter, cream, and so on)	**Fatty acids and glycerol**
Pancreatic juice Protease (trypsin)†	Proteins (either intact or partially digested)	Proteoses, peptides, and **amino acids**
Lipase (steapsin)	Bile—emulsified fats	**Fatty acids and glycerol**
Amylase (amylopsin)	Starch	Maltose
Intestinal juice (succus entericus) Peptidases	Peptides	**Amino acids**
Sucrase	Sucrose (cane sugar)	**Glucose and fructose‡** (simple sugars or monosaccharides)
Lactase	Lactose (milk sugar)	**Glucose and galactose** (simple sugars)
Maltase	Maltose (milk sugar)	**Glucose** (grape sugar)

*Substances in boldface type are end products of digestion; that is, completely digested foods ready for absorption.
†Secreted in inactive form (trypsinogen); activated by enterokinase, an enzyme in the intestinal juice.
‡Glucose is also called dextrose; fructose is called levulose.

depend on the absorption of digested food and its transportation to them by the circulating blood. On p. 178 we mentioned something about the structure of the intestinal lining that enables it to absorb foods rapidly. Do you recall what this is?

Metabolism

A good phrase to remember in connection with the word metabolism is "use of foods," for basically this is what metabolism is—the use the body makes of foods once they have been digested, absorbed, and circulated to cells. It uses them in two ways: as an energy source and as building blocks for making complex chemical compounds. Before they can be used in these two ways, foods have to enter cells and there undergo many chemical changes. All chemical reactions that release energy from food molecules together make up the process of catabolsum—a vital process, since it is the only way that the body has of supplying itself with energy for doing any of its many kinds of work. The many chemical reactions that build food molecules into more complex chemical compounds together constitute the process of anabolism. Catabolism and anabolism together make up the process of metabolism.

Carbohydrate metabolism

Carbohydrates are the body's preferred energy food. Human cells catabolize glucose rather than other substances as long as enough glucose enters them to supply their energy needs. Two series of chemical reactions, occurring in a precise sequence, make up the process of glucose metabolism. *Glycolysis* is the name given the first series of reactions; *citric acid cycle* is the name of the second series. Glycolysis, as Fig. 12-7 shows, changes glucose to pyruvic acid, and then the citric acid cycle changes the

pyruvic acid to carbon dioxide. Glycolysis takes place in the cytoplasm of a cell, whereas the citric acid cycle goes on in the mitochondria, the cell's miniature power plants. Glycolysis uses no oxygen; it is an anaerobic process. The citric acid cycle, in contrast, is an oxygen-using or aerobic process.

While the chemical reactions of glycolysis and the citric acid cycle are going on, energy stored in the glucose molecule is being released. Almost instantaneously, however, more than half of it is put back into storage, not in glucose molecules but in the molecules of another compound, ATP (adenosine triphosphate). The rest of the energy originally stored in the glucose molecule is released as heat. ATP serves as the direct source of energy for doing cellular work in all kinds of living organisms from one-cell plants to billion-cell animals, including man. Among biological compounds, there-

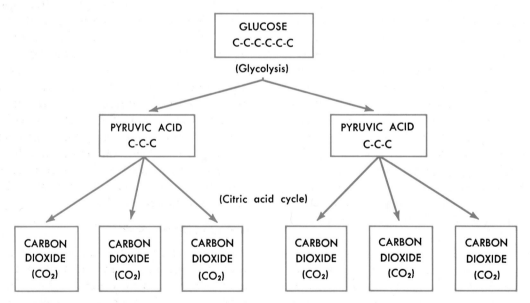

Fig. 12-7 Scheme to show that catabolism breaks larger molecules down into smaller ones. Glycolysis splits one molecule of glucose (six carbon atoms) into two molecules of pyruvic acid (three carbon atoms each). The citric acid cycle converts each pyruvic acid molecule to three carbon dioxide molecules.

fore, ATP ranks as one of the most important. The energy stored in ATP molecules differs in two ways from the energy stored in food molecules. The energy in ATP molecules can be released almost instantaneously, and it can be used directly to do cellular work. Release of energy from food molecules occurs much more slowly because it accompanies the long series of chemical reactions that make up the process of catabolism. For some reason, energy released from food molecules cannot be used directly for doing cellular work. It must first be transferred to ATP molecules and be released explosively from them.

Although glucose is mainly catabolized, small amounts of it are anabolized. One process of glucose anabolism is called glycogenesis. Carried on chiefly by liver and muscle cells, *glycogenesis* consists of a series of reactions that join glucose molecules together, like so many beads in a necklace, to form glycogen, a compound sometimes called "animal starch."

Something worth noticing is that the amount of food in the blood normally does not change very much, not even when we go without food for many hours, when we exercise and use a lot of food for energy, or when we sleep and use little food for energy. The amount of glucose, for example, usually stays at about 80 to 120 milligrams in 100 milliliters of blood.

Several hormones help regulate carbohydrate metabolism to keep blood glucose normal. *Insulin* is one of the most important of these. It acts in some way not yet definitely known to make glucose leave the blood and enter the cells at a more rapid rate. As insulin secretion increases, more glucose leaves the blood and enters the cells. The amount of glucose in the blood therefore tends to decrease while the rate of glucose metabolism in cells tends to increase. Too little insulin secretion, such as

occurs in diabetes mellitus, produces the opposite effects. Less glucose leaves the blood and enters the cells; more glucose therefore remains in the blood and less glucose is metabolized by cells. In other words, high blood glucose (hyperglycemia) and a low rate of glucose metabolism characterize insulin deficiency. Insulin is the only hormone that functions to lower the blood glucose level. Several other hormones, on the other hand, tend to increase it. Growth hormone secreted by the anterior pituitary gland, hydrocortisone secreted by the adrenal cortex, and epinephrine secreted by the adrenal medulla are three of the most important hormones that tend to increase blood glucose. More information about these hormones and others that help control metabolism appears on pp. 116, 122, and 124.

Fat metabolism

Fats, like carbohydrates, are primarily energy foods. If cells have inadequate amounts of glucose to catabolize, they immediately shift to the catabolism of fats for their energy supply. This happens normally whenever a person goes without food for many hours. It happens abnormally in diabetic individuals. Because of an insulin deficiency, too little glucose enters the cells of a diabetic person to supply all of his energy needs. Result? The cells catabolize fats to make up the difference. In all persons, fats not needed for catabolism are anabolized and stored in adipose tissue.

Protein metabolism

Protein catabolism occurs to some extent, but more important is protein anabolism, the process by which the body builds amino acids into complex protein compounds—for example, enzymes and proteins that form the structural part of the cell.

Metabolic rates

The *basal metabolic rate* (BMR) is the rate at which food is catabolized under basal conditions—that is, when the individual is awake, is not digesting food, and is not adjusting to a cold external temperature. Or stated differently, the basal metabolic rate is the number of calories of heat that must be produced per day by catabolism just to keep the body alive, awake, and comfortably warm. For the determination of the basal metabolic rate, the amount of oxygen the individual inhales in a specific length of time is measured, and this figure is used to calculate the basal metabolic rate, stated in calories.

The basal metabolic rate is an indirect measure of thyroid gland functioning. So, too, is the amount of protein-bound iodine (PBI) in venous blood. A higher than normal concentration of PBI indicates higher than normal secretion of thyroid hormone and higher than normal metabolic rate. PBI determination has replaced BMR determination in clinical use because PBI determination is a simpler procedure.

A person's BMR represents the amount of food that his body must catabolize each day for him simply to stay alive and awake in a comfortably warm environment. To provide energy for him to do muscular work and to digest and absorb any food, an additional amount of food must be catabolized. How much more food must be catabolized depends mainly on how much work the individual does. The more active he is, the more food his body must catabolize and the higher his total metabolic rate will be. The *total metabolic rate* (TMR) is the total amount of energy (expressed in calories) used by the body per day (Fig. 12-8).

When the number of calories in your daily food intake equals your TMR, your weight remains constant (except for possible variations resulting from water retention or water loss). When your daily food contains more calories than your TMR, you gain weight; when your daily food contains fewer calories than your TMR, you lose weight. These weight control principles never fail to operate. Nature never forgets to count calories. Reducing diets make use of this knowledge. They contain fewer calories than the TMR of the individual eating the diet.

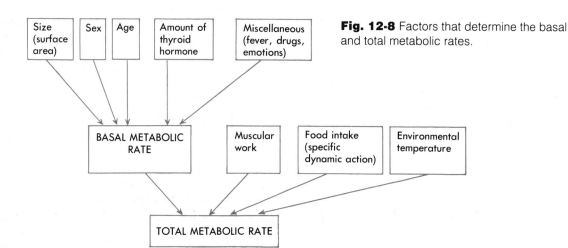

Fig. 12-8 Factors that determine the basal and total metabolic rates.

Outline summary

Mouth

A Roof—formed by hard palate (parts of maxillary and palatine bones) and soft palate, an arch-shaped muscle separating mouth from nasopharynx; uvula, a downward projection of soft palate

B Floor—formed by tongue and its muscles; papillae, small elevations on mucosa of tongue; taste buds, found in many papillae; frenulum, fold of mucous membrane that helps anchor tongue to floor of mouth

Teeth

A Twenty teeth in temporary set; average age for cutting first tooth about 6 months; set complete at about 2 years of age

B Thirty-two teeth in permanent set; 6 years about average age for starting to cut first permanent tooth; set complete usually between ages of 17 and 24 years

C Names of teeth—see Fig. 12-2

D Structures of tooth—see Fig. 12-3

Salivary glands

A Parotid glands

B Submandibular glands

C Sublingual glands

Pharynx

Esophagus

Stomach

A Size—expands after large meal; about size of large sausage when empty

B Pylorus—lower part of stomach; pyloric sphincter muscle closes opening of pylorus into duodenum

C Wall—many smooth muscle fibers; contractions produce churning movements and peristalsis

D Lining—mucous membrane; many microscopic glands that secrete gastric juice and hydrochloric acid into stomach; mucous membrane lies in folds (rugae) when stomach is empty

Small intestine

A Size—about 20 feet long but only an inch or so in diameter

B Divisions

   1 Duodenum

   2 Jejunum

   3 Ileum

C Wall—contains smooth muscle fibers that contract to produce peristalsis

D Lining—mucous membrane; many microscopic glands (intestinal glands) secrete intestinal juice; villi (microscopic finger-shaped projections from surface of mucosa into intestinal cavity) contain blood and lymph capillaries

Liver and gallbladder

A Size and location—largest gland; fills upper right section of abdominal cavity and extends over into left side

B Functions—secretes bile; helps maintain normal blood glucose by carrying on processes of glycogenesis, glycogenolysis, and gluconeogenesis; forms prothrombin, fibrinogen, and certain other blood proteins; also performs several other functions

C Ducts

   1 Hepatic—drains bile from liver

   2 Cystic—duct by which bile enters and leaves gallbladder

   3 Common bile—formed by union of hepatic and cystic ducts; drains bile from hepatic or cystic ducts into duodenum

Pancreas

A Location—behind stomach

B Functions

   1 Pancreatic cells secrete pancreatic juice into pancreatic ducts; main duct empties into duodenum

   2 Islands of Langerhans—cells not connected with pancreatic ducts; beta cells secrete insulin and alpha cells secrete glucagon into blood

Large intestine

A Divisions

   1 Cecum

   2 Colon—ascending, transverse, descending, and sigmoid

   3 Rectum

B Opening to exterior—anus

C Wall—contains smooth muscle fibers that contract to produce churning, peristalsis, and defecation

D Lining—mucous membrane

Appendix

Blind tube off cecum; no known functions in humans

Peritoneum

A Definitions—peritoneum, serous membrane lining abdominal cavity and covering abdominal organs; parietal layer of peritoneum lines abdominal cavity; visceral layer of peritoneum covers abdominal organs; peritoneal space lies between parietal and visceral layers

B Extensions—largest ones are the mesentery and greater omentum; mesentery is extension of parietal peritoneum, which attaches most of small intestine to posterior abdominal wall; greater omentum, or "lace apron," hangs down from lower edge of stomach and transverse colon over intestines

Digestion

Meaning—changing foods so that they can be absorbed and used by cells

A Mechanical digestion—chewing, swallowing, and peristalsis break food into tiny particles, mix them well with the digestive juices, and move them along the digestive tract

B Chemical digestion—breaks up large food molecules into compounds having smaller molecules; brought about by digestive enzymes

C Protein digestion—starts in stomach; completed in small intestine
  1 Gastric juice enzymes, rennin and pepsin, partially digest proteins
  2 Pancreatic enzyme, trypsin, completes digestion of proteins to amino acids
  3 Intestinal enzyme, erepsin, completes digestion of partially digested proteins to amino acids

D Carbohydrate digestion—mainly in small intestine
  1 Pancreatic enzyme, amylopsin—changes starches to maltose
  2 Intestinal juice enzymes
    a Maltase changes maltose to glucose
    b Sucrase changes sucrose to glucose
    c Lactase changes lactose to glucose

E Fat digestion
  1 Gastric lipase changes small amount of fat to fatty acids and glycerin in stomach
  2 Bile contains no enzymes but emulsifies fats (breaks fat droplets into very small droplets)
  3 Pancreatic lipase changes emulsified fats to fatty acids and glycerin in small intestine

Absorption

A Meaning—digested food moves from intestine into blood or lymph

B Where absorption occurs—foods and water from small intestine; water also absorbed from large intestine

Metabolism

A Meaning—use of foods by body cells for energy and building complex compounds

B Catabolism—breaks food molecules down into carbon dioxide and water, releasing their stored energy; oxygen used up in catabolism

C Anabolism—builds food molecules into complex substances

D Carbohydrates primarily catabolized for energy but small amounts are anabolized by glycogenesis (a series of chemical reactions that changes glucose to glycogen—occurs mainly in liver cells where glycogen is stored); glycogenolysis is process (series of chemical reactions) by which glycogen is changed back to glucose

E Blood glucose (imprecisely, blood sugar)—normally stays between about 80 and 120 milligrams per 100 milliters of blood; *insulin* accelerates movement of glucose out of blood into cells, therefore tends to decrease blood glucose and increase glucose catabolism

F Fats both catabolized to yield energy and anabolized to form adipose tissue

G Proteins primarily anabolized and secondarily catabolized

H Metabolic rates
  1 Basal metabolic rate (BMR)—rate of metabolism when person is lying down, but awake, when about 12 hours have passed since last meal, and when environment is comfortably warm
  2 Total metabolic rate (TMR)—the total amount of energy, expressed in calories, used by the body per day
  3 Protein-bound iodine (PBI)—indirect measure of thyroid secretion and of metabolic rate

New words

absorption	glycogenesis
ascites	glycogenolysis
digestion	peristalsis
gluconeogenesis	peritoneum

Review questions

1 What organs form the gastrointestinal tract?
2 Identify each of the following structures: jejunum, cecum, colon, duodenum, and ileum.
3 If you inserted 9 inches of an enema tube through the anus, the tip of the tube would probably be in what structure?
4 How many teeth should an adult have?
5 How many teeth should a child 2½ years old have? Would he have some of each of the following teeth: incisors, canines, bicuspids, and tricuspids? If not, which ones would he not have?
6 Identify each of the following:

   islands of Langerhans   rugae
   parotid glands   villi
   pylorus

7 What process do liver cells carry on that prevents the level of blood glucose from becoming dangerously high after a heavy meal?
8 What two processes do liver cells carry on that prevent a dangerously low blood glucose level from developing during fasting and between meals?
9 In what organ does the digestion of starches begin?
10 What digestive juice contains no enzymes?
11 Only one digestive juice contains enzymes for digesting all three kinds of food. Which juice is this? In what organ does it do its work?
12 What kind of food is not digested in the stomach?
13 Which digestive juice emulsifies fats?
14 What three digestive juices act on foods in the small intestine?
15 What juices digest carbohydrates? proteins? fats?
16 Explain as briefly and clearly as you can what each of the following terms means:

   absorption   digestion
   anabolism   metabolism
   catabolism

17 Explain why you think the following statement is true or false: "If you do not want to gain or lose weight but just stay the same, you must eat just enough food to supply the calories of your BMR. If you eat more than this, you will gain; if you eat less than this, you will lose."

chapter 13

The respiratory system

No one needs to be told how important his respiratory system is. The respiratory system serves the body much as a lifeline to an oxygen tank serves a deep-sea diver. Think how panicky you would feel if suddenly your lifeline became blocked—if you could not breathe for a few seconds! Of all the substances that cells and therefore the body as a whole must have to survive, oxygen is by far the most crucial. A person can live a few weeks without food, a few days without water, but only a few minutes without oxygen. Constant removal of carbon dioxide from the body is just as important for survival as a constant supply of oxygen. The organs of the respiratory system are designed to perform two basic functions—they serve as (1) an *air distributor* and (2) a *gas exchanger* for the body. The respiratory system ensures that oxygen is supplied to and carbon dioxide is removed from the body's cells. In this chapter the structural plan of the respiratory system will be considered first, then the respiratory organs will be discussed individually, and finally some facts about respiration that are important for a nurse or allied health professional to know will be given.

Structural plan

The respiratory organs are the nose, pharynx, larynx, trachea, bronchi, and lungs. Their basic structural design is that of a tube with many branches ending in millions of extremely tiny, extremely thin-walled sacs called *alveoli*. A network of capillaries fits like a tight-fitting hairnet around each microscopic alveolus. Incidentally, this is a good place for us to think again about a principle already mentioned several times, namely, that structure determines function. The function of alveoli—in fact, the function of the entire respiratory system—is to distribute air close enough to blood for a gas exchange to take place between air and blood. Two facts about the structure of alveoli make them able to perform this function admirably. First, the wall of each alveolus is made up of a single layer of cells and so are the walls of the capillaries around it. This means that between the blood in the capillaries and the air in the alveolus there is a barrier probably less than $1/5,000$ of an inch thick! This extremely thin barrier is called the *respiratory membrane*. Note in Fig. 13-2 that the surface of the respiratory membrane inside the alveolus is covered by a substance called *surfactant*. This important substance helps to reduce surface tension in the alveoli and keep them from collapsing as air moves in and out during respiration. Second, there are millions of alveoli. This means that together they make an enormous surface (in the neighborhood of 1,100 square feet, an area many times larger than the surface of the entire body) where larger amounts of oxygen and carbon dioxide can rapidly be exchanged.

Nose

Air enters the respiratory tube through the nostrils (nares) into the right and left nasal cavities. A partition, the nasal septum, separates these two cavities; mucous membrane lines them. The surface of the nasal cavities is moist from mucus and warm from blood flowing just under it. Nerve endings responsible for the sense of smell are located in the nasal mucosa. Four paranasal sinuses (frontal, maxillary, sphenoidal, and ethmoidal) drain into the nasal cavities. Because the mucosa that lines the sinuses is continuous with the mucosa that lines the nose, sinus infections often develop from colds in which the nasal mucosa is inflamed.

Note in Fig. 13-3 that three shelflike structures called *conchae* protrude into the nasal cavity on each side. Think of the mucosa-covered conchae as partitions that greatly increase the surface area over which air must flow as it passes through the nasal cavity. As air moves over the conchae and through the nasal cavities, it is warmed and humidified. In addition, the presence of mucus helps trap dust particles that might otherwise irritate the delicate lining of the respiratory passageways.

Pharynx

The pharynx is the structure that many of us call the throat. Although only about 12.5

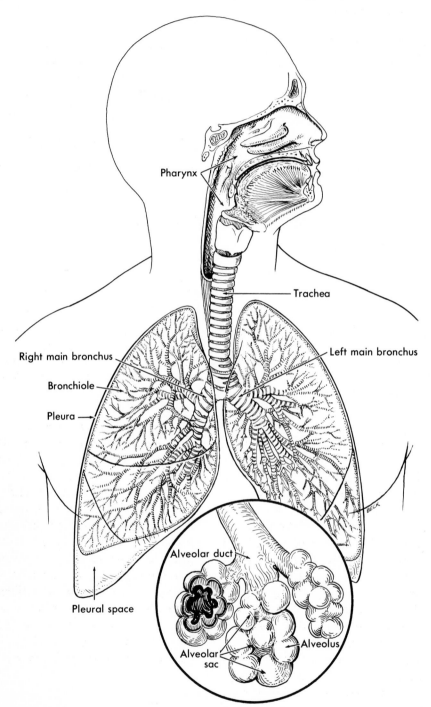

Fig. 13-1 Structural plan of the respiratory organs showing the pharynx, trachea, bronchi, and lungs. The inset shows the grapelike alveolar sacs where the interchange of oxygen and carbon dioxide takes place through the thin walls of the alveoli. Capillaries (not shown) surround the alveoli.

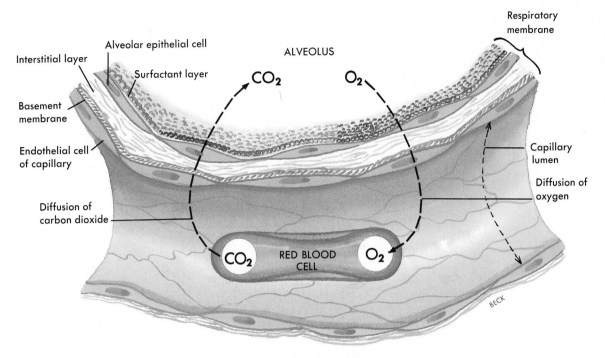

Fig. 13-2 A small portion of the respiratory membrane greatly magnified. An extremely thin interstitial layer of tissue separates the endothelial cell and basement membrane on the capillary side from the the epithelial cell and surfactant layer on the alveolar side of the respiratory membrane. The total thickness of the respiratory membrane is less than 1/5,000 of an inch!

centimeters (5 inches) long, it can be divided into three portions. The uppermost part of the tube just behind the nasal cavities is called the *nasopharynx*. The portion behind the mouth is call the *oropharynx*. The last or lowest segment is called the *laryngopharynx*. The pharynx as a whole serves the same purpose for the respiratory and digestive tracts as a hallway serves for a house. Air and food pass through the pharynx on their way to the lungs and the stomach, respectively. Air enters the pharynx from the two nasal cavities and leaves it by way of the larynx; food enters it from the mouth and leaves it by way of the esophagus. The right and left eustachian (auditory) tubes open into the nasopharynx; they connect each middle ear with the nasopharynx (Fig. 13-3). The lining of the eustachian tubes is continuous with the lining of both the nasopharynx and middle ear. There-

fore, just as sinus infections can develop from colds in which the nasal mucosa is inflamed, middle ear infections can develop as a result of inflammation of the nasopharynx. Two pairs of organs that seem to give more trouble than service to the body (the *tonsils* and *adenoids*) are also located in the pharynx. The tonsils are found in the oropharynx. The adenoids are located in the nasopharynx. If the adenoids become enlarged, they can make it difficult or impossible for air to travel from the nose into the throat. When this happens, the individual keeps his mouth open to breathe and is described as appearing "adenoidal."

Larynx

The larynx, or voice box, is located just below the pharynx. It is composed of several pieces of cartilage. You know the larg-

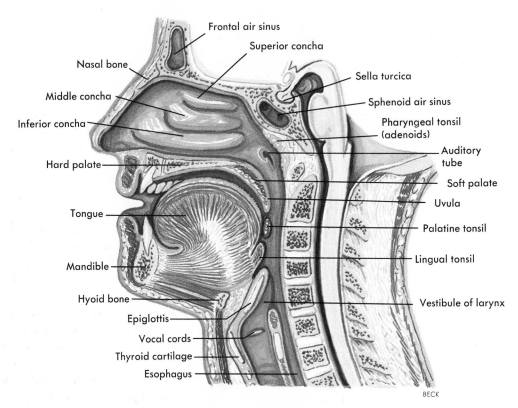

Fig. 13-3 Sagittal section through the face and neck. The nasal septum has been removed, exposing the lateral wall of the nasal cavity. Note the position of the conchae.

est of these (the *thyroid cartilage*) as the "Adam's apple."

Two short fibrous bands called the vocal cords stretch across the interior of the larynx. Muscles that attach to the cartilages of the larynx can pull on these cords in such a way that they become either tense and short or relaxed and long. When they are tense and short, the voice sounds high pitched; when they are relaxed and long, it sounds low pitched.

Trachea

The trachea or windpipe is a tube about 11 centimeters (4½ inches) long that extends from the larynx in the neck to the bronchi in the chest cavity (Figs. 13-1 and 13-4). The trachea performs a simple but vital function—it furnishes part of the open passageway through which air can reach the lungs from the outside.

By pushing with your fingers against your throat about an inch above the sternum, you can feel the shape of the trachea or windpipe. Only if you use considerable force can you squeeze it closed. Nature has taken precautions to keep this lifeline open. Its framework is made of an almost noncollapsible material—15 or 20 C-shaped rings of cartilage placed one above the other with only a little soft tissue between them. Despite this structural safeguard, closing of the trachea does sometimes occur. A tumor or an infection may enlarge the lymph nodes of the neck so much that they squeeze the trachea shut, or a person may aspirate (breathe in) a piece of food or something else that blocks the windpipe. Since air has no other way to get to the lungs, complete tracheal obstruction causes death in a matter of minutes.

Bronchi, bronchioles, and alveoli

One way to picture the thousands of air tubes that make up the lungs is to think of an upside-down tree. The trachea is the main trunk of this tree; the right bronchus (the tube leading into the right lung) and the left bronchus (the tube leading into the left lung) are the trachea's first branches. In each lung they branch into smaller bronchi, which branch into bronchioles. The smallest bronchioles end in structures shaped like miniature bunches of grapes (Fig. 13-1). The smallest bronchioles subdivide into microscopic-sized tubes called *alveolar ducts*, which resemble the main stem of a bunch of grapes. Each alveolar duct ends in several *alveolar sacs*, each of which resembles a cluster of grapes, and the wall of each alveolar sac is made up of numerous *alveoli*, each of which resembles a single grape. How well alveoli are suited to their function of exchanging gases between air and blood has been mentioned. They can perform this function effectively mainly because alveoli are extremely thin walled, each alveolus lies in contact with a blood capillary, and there are millions of alveoli in each lung.

Because the air tubes of the bronchial tree are embedded in connective tissue, you cannot see them by looking at the lungs from the outside. But if you cut into the lungs, the bronchi and bronchioles show up clearly.

Lungs and pleura

The lungs are fairly large organs. As Fig. 13-4 shows, they fill the entire chest cavity (all but the space in the center occupied mainly by the heart and large blood vessels). The narrow part of each lung, up under the collarbone, is its *apex;* the broad lower part, resting on the diaphragm, is its *base*. The *pleura* covers the outer surface of the lungs and lines the inner surface of the rib cage. The pleura resembles the peritoneum in structure and function but differs from it in location. Both are extensive, thin, moist, slippery membranes. Both line a large, closed cavity of the body and cover organs located in them. The *parietal pleura* lines the thoracic cavity, the *visceral pleura* covers the lungs, and the *pleural space* lies between the two pleural membranes. Where are the parietal peritoneum, the visceral peritoneum, and the peritoneal space located? (See p. 181 if you do not know.)

Normally, the pleural space contains just enough fluid to make both portions of the pleura moist and slippery and able to glide easily against each other as the lungs expand and deflate with each breath. However, the pleural space sometimes becomes distended with a large amount of fluid. The extra fluid presses on the lungs and makes it hard for the patient to breathe. In this case the physician may decide to remove the excess pleural fluid by means of a hollow, tubelike instrument that he pushes through the patient's chest wall into the pleural space. If you were to ask the patient what the doctor did to him, he might answer, "He tapped my chest." If you asked the doctor the technical name of the procedure, he would answer, "A *thoracentesis*."

Pleurisy and pneumothorax are other terms having to do with pleura and pleural space. *Pleurisy* (or pleuritis) is an inflammation of the pleura. *Pneumothorax* is the

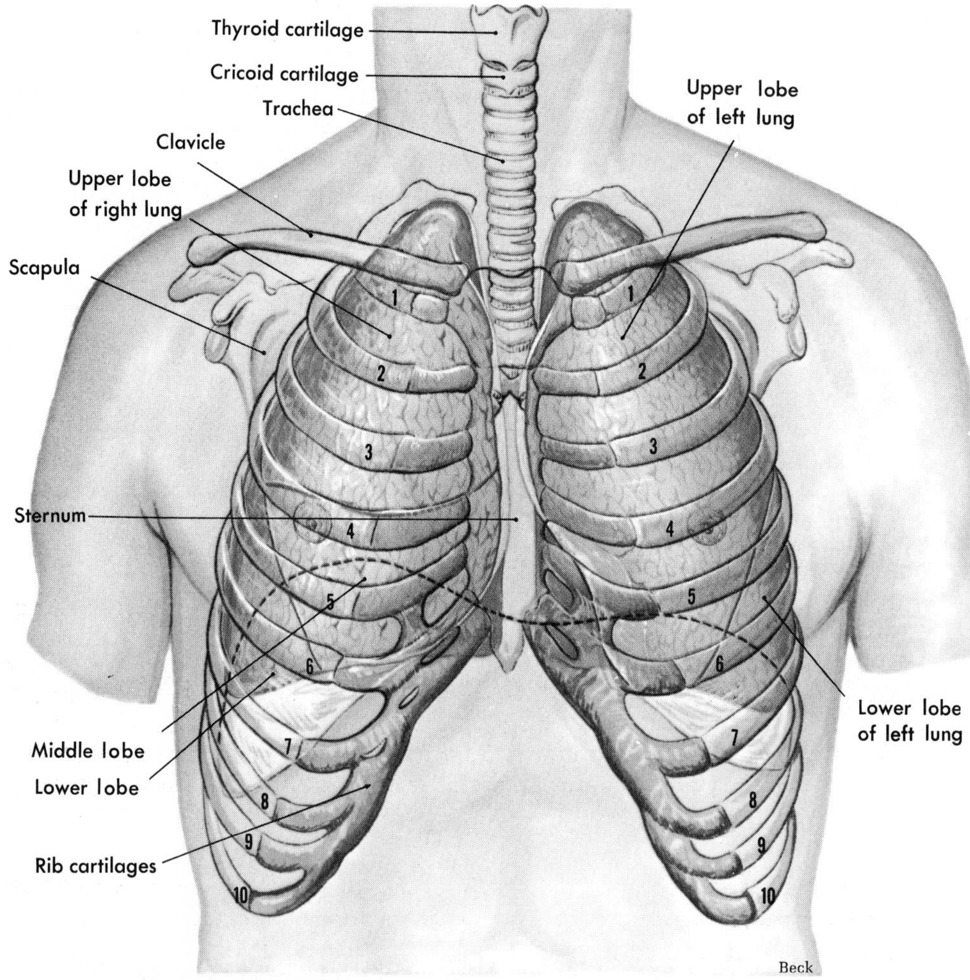

Fig. 13-4 Projection of the lungs and trachea in relation to rib cage and clavicles. Dotted line shows location of dome-shaped diaphragm at the end of expiration and before inspiration. Note that apex of each lung projects above the clavicle. Ribs 11 and 12 are not visible in this view.

presence of air in the pleural space of one side of the chest. The additional air increases the pressure on the lung on that side and causes it to collapse. While collapsed, the lung does not function in breathing.

Respiration

Respiration means exchange of gases (oxygen and carbon dioxide) between a living organism and its environment. If the organ-

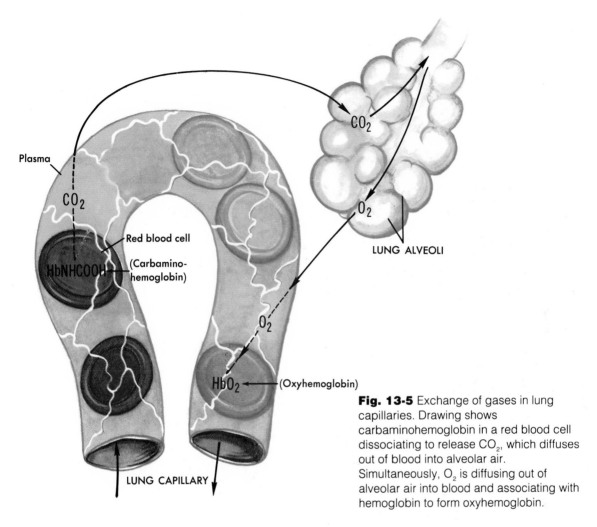

Fig. 13-5 Exchange of gases in lung capillaries. Drawing shows carbaminohemoglobin in a red blood cell dissociating to release CO_2, which diffuses out of blood into alveolar air. Simultaneously, O_2 is diffusing out of alveolar air into blood and associating with hemoglobin to form oxyhemoglobin.

ism consists of only one cell, gases can move directly between it and the environment. If, however, the organism consists of billions of cells, as do our bodies, then most of its cells are too far from the air for a direct exchange of gases. To overcome this difficulty, a pair of organs—the lungs—is provided where air and a circulating fluid (blood) can come close enough to each other for oxygen to move out of the air into blood while carbon dioxide moves out of the blood into air. Breathing, or *pulmonary ventilation*, is the process that moves air into and out of the lungs. It is the process that makes possible the exchange of gases be-

tween air and blood and eventually between blood and cells.

Exchange of gases in lungs

As blood flows through the thousands of tiny lung capillaries, carbon dioxide leaves it and oxygen enters it. This two-way exchange of gases between the blood in the lung capillaries and the air in the alveoli comes about in the following way. Venous blood enters lung capillaries. (These are the only capillaries where this is true; arterial blood enters all other capillaries—"tissue capillaries" as they are called. As you can see in Fig. 13-5, the hemoglobin inside the red blood cells in venous blood is combined with carbon dioxide rather than with oxygen and so is called carbaminohemoglobin. As the blood flows along through the lung capillaries, carbaminohemoglobin breaks down into carbon dioxide and hemoglobin. Carbon dioxide molecules move out of red blood cells into the plasma and pass quickly through the thin capillary-alveolar membrane into the alveoli. Then from the alveoli, carbon dioxide leaves the body in the expired air. As carbon dioxide molecules diffuse out of the lungs into the atmosphere, oxygen molecules are diffusing from the lungs' alveoli into the blood plasma and on into the red blood cells. Here they combine with hemoglobin to form oxyhemoglobin.

The exchange of gases between lung capillary blood and alveolar air (carbon dioxide out of the blood, oxygen into the blood) changes venous blood to arterial blood. Fig. 13-5 shows these changes. Arterial blood contains more oxygen and less carbon dioxide than does venous blood.

Exchange of gases in tissues

The exchange of gases between the blood in tissue capillaries and the cells that make up the tissues is just the opposite of the exchange of gases between the blood in lung capillaries and the air in alveoli. As shown in Fig. 13-6, in the tissue capillaries oxyhemoglobin breaks down into oxygen and hemoglobin. Oxygen molecules move rapidly out of the blood through the tissue capillary membrane into the interstitial fluid and on into the cells that compose the tissues. While this is happening, carbon dioxide molecules are leaving the cells, entering the tissue capillaries, and uniting with hemoglobin molecules to form carbaminohemoglobin. In other words, arterial blood enters tissue capillaries and is changed into venous blood as it flows through them.

Mechanics of breathing

Breathing involves not only the organs of the respiratory system but also the brain, spinal cord, nerves, certain skeletal muscles, and even some bones and joints. In breathing, nerve impulses stimulate the diaphragm to contract, and, as it contracts, its shape changes. Its domelike shape of the diaphragm flattens out. Then, instead of protruding up into the chest cavity, it moves down toward the abdominal cavity. Thus the contraction or flattening of the diaphragm makes the chest cavity longer from top to bottom. Other muscle contractions raise the rib cage to make the chest cavity wider and greater in depth from front to back. As the chest cavity enlarges, the lungs expand along with it and air rushes into them and down into the alveoli. This part of respiration is called inspiration. For expiration to take place, the diaphragm and other respiratory muscles relax, making the chest cavity smaller and thereby squeezing air out of the lungs.

Volumes of air exchanged in pulmonary ventilation

A special device called a *spirometer* is used to measure the amount of air exchanged in breathing. Fig. 13-7 illustrates the various pulmonary volumes, which can be measured as a subject breathes into a spirometer. Ordinarily we take 500 milliliters (about a pint) of air into our lungs. Because this amount comes and goes regularly like the tides of the sea, it is referred to as the *tidal volume* (TV). The largest amount of air that we can breathe in and out in one inspiration and expiration is known as the *vital capacity* (VC). In normal young men this is about 4,800 milliliters. Tidal volume and vital capacity are frequently measured in patients with lung or heart disease, conditions that often lead to abnormal volumes of air being moved in and out of the lungs.

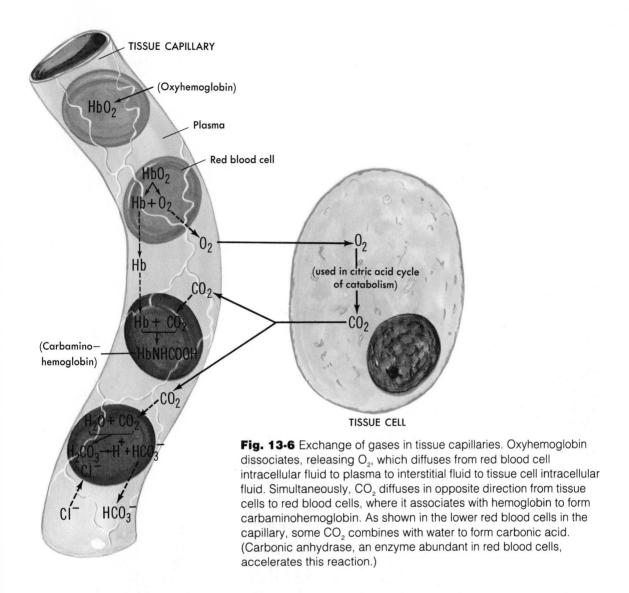

TISSUE CAPILLARY

(Oxyhemoglobin)

HbO_2

Plasma

Red blood cell

HbO_2

$Hb + O_2$

O_2

Hb

O_2

(used in citric acid cycle of catabolism)

CO_2

$Hb + CO_2$

CO_2

(Carbamino— hemoglobin)

HbNHCOOH

CO_2

$H_2O + CO_2$

$H_2CO_3 \rightarrow H^+ + HCO_3^-$

Cl^-

Cl^- HCO_3^-

TISSUE CELL

Fig. 13-6 Exchange of gases in tissue capillaries. Oxyhemoglobin dissociates, releasing O_2, which diffuses from red blood cell intracellular fluid to plasma to interstitial fluid to tissue cell intracellular fluid. Simultaneously, CO_2 diffuses in opposite direction from tissue cells to red blood cells, where it associates with hemoglobin to form carbaminohemoglobin. As shown in the lower red blood cells in the capillary, some CO_2 combines with water to form carbonic acid. (Carbonic anhydrase, an enzyme abundant in red blood cells, accelerates this reaction.)

Observe the area in Fig. 13-7 that represents the *expiratory reserve volume* (ERV). This is the amount of air that can be forcibly exhaled after expiring the tidal volume. Compare this with the area in Fig. 13-7 that represents the *inspiratory reserve volume* (IRV). The IRV is the amount of air that can be forcibly inspired over and above a normal inspiration. As the tidal volume increases, both the ERV and IRV will decrease. Note in Fig. 13-7 that vital capacity is the total of tidal volume, inspiratory reserve volume, and expiratory reserve volume. Or expressed in another way: $VC = TV + IRV + ERV$. *Residual volume* (RV) is simply the air that remains in the lungs after the most forceful expiration.

By now you know that the body uses oxygen to obtain energy for the work it has to do. Briefly, energy is stored in foods and

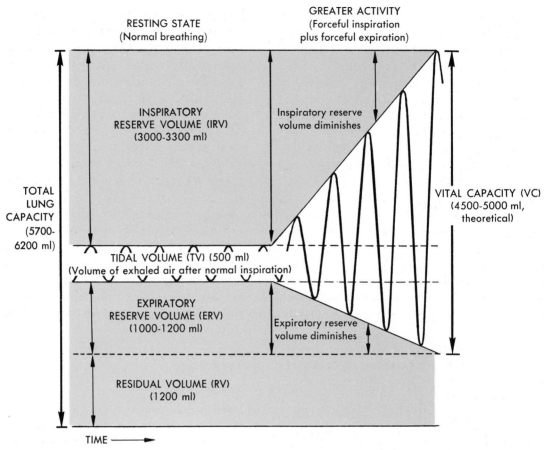

Fig. 13-7 During normal, quiet respirations, the atmosphere and lungs exchange about 500 ml of air (tidal volume). With a forcible inspiration, about 3,300 ml more air can be inhaled (inspiratory reserve volume). After a normal inspiration and normal expiration, approximately 1,000 ml more air can be forcibly expired (expiratory reserve volume). Vital capacity is the largest amount of air that can enter and leave the lungs during respiration. Residual volume is the air that remains trapped in the alveoli.

the oxidation reactions of catabolism make this energy available for all kinds of cellular work (pp. 184 and 185). Therefore the more work the body does, the more oxygen must be delivered to its millions of cells. One way this is accomplished is by increasing the rate and depth of respirations. Although we may take only sixteen breaths a minute when we are not moving about, when we

are exercising we take considerably more than this. Not only do we take more breaths; we also breathe in more air with each breath. Instead of about a pint of air, we may breathe deeply enough to take in several pints—sometimes even up to the limit of our vital capacity.

The way respirations are made to increase during exercise is an interesting example of the body's automatic regulation of its vital functions. When we are exercising, cells carry on catabolism at a faster rate than usual. This increased metabolic rate means that more carbon dioxide is formed in the cells and that more carbon dioxide enters the blood in the tissue capillaries and increases venous blood's carbon dioxide concentration. This higher carbon dioxide concentration has a stimulating effect on the neurons of the respiratory center of the medulla. They send out more impulses to the respiratory muscles, and as a result, respiration increases, becoming both faster and deeper. Therefore, as more air moves in and out of the lungs per minute, more carbon dioxide leaves the blood, more oxygen enters it, and more energy becomes available for doing the extra work of exercise.

To help supply cells with more oxygen when they are doing more work, automatic adjustments occur not only in respirations, but also in circulation. Most notably, the heart beats faster and harder and therefore pumps more blood through the body each minute. This means that the millions of red blood cells make more round trips between the lungs and tissues each minute and so deliver more oxygen per minute to tissue cells.

Disorders of respiratory system

Many different kinds of diseases and injuries may be responsible for respiratory disorders. For example, tuberculosis and lung cancer may destroy part of the lungs and pneumonia may plug up alevoli. A brain hemorrhage may depress the respiratory center causing respiratiosn to become slow and labored or even stop completely. In recent years a disease rarely heard of in our grandfathers' generation—*emphysema*—

has become increasingly common. In this disease the walls of many of the alveoli become greatly overstretched. As the alveoli enlarge, their walls rupture and then fuse into large irregular spaces. Air becomes trapped in the lungs. As a result, less air than normal is exhaled and inhaled, and less oxygen and carbon dioxide are exchanged between alveolar air and blood. Emphysema victims therefore develop *hypoxia*. A synonym for hypoxia is oxygen deficiency. If a person is hypoxic, the cells receive less oxygen than what is needed for normal functioning. Lung disease is not the only abnormality that can produce hypoxia. Anemia, for example, can also cause hypoxia. An anemic individual's red blood cells contain less hemoglobin than normal and so they transport less oxygen than normal.

Outline summary

Structural plan

Basic plan of respiratory system similar to an inverted tree if it were hollow; leaves of tree would be comparable to alveoli, with the microscopic sacs enclosed by networks of capillaries

Nose

A Structure
  1 Nasal septum separates interior of nose into two cavities
  2 Mucous membrane lines nose
  3 Frontal, maxillary, sphenoidal, and ethmoidal sinuses drain into nose
B Functions
  1 Warms and moistens air inhaled
  2 Contains sense organs of smell

Pharynx

A Structure
  1 Pharynx (throat) about 12.5 centimeter (5 inches) long
  2 Divided into nasopharynx, oropharynx, and laryngopharynx
  3 Two nasal cavities, mouth, esophagus,

larynx, and eustachian tubes all have openings into pharynx

  4 Adenoids and openings of eustachian tubes open into nasopharynx; tonsils found in oropharynx

  5 Mucous membrane lines pharynx

B Functions

  1 Passageway for food and liquids

  2 Air distribution; passageway for air

Larynx (voice box)

A Structure

  1 Several pieces of cartilage form framework

  2 Mucous lining

  3 Vocal cords stretch across interior of larynx

B Functions

  1 Air distribution; passageway for air to move to and from lungs

  2 Voice production

Trachea (windpipe)

A Structure

  1 Tube about 11 centimeters ($4\frac{1}{2}$ inches) long that extends from larynx into the thoracic cavity

  2 Mucous lining

  3 C-shaped rings of cartilage hold trachea open

B Function—passageway for air to move to and from lungs

Bronchi, bronchioles, and alveoli

A Structure

  1 Trachea branches into right and left bronchi

  2 Each bronchus branches into smaller and smaller tubes called bronchioles

  3 Bronchioles end in clusters of microscopic alveolar sacs, the walls of which are made up of alveoli

B Function

  1 Bronchi and bronchioles—air distribution; passageway for air to move to and from alveoli

  2 Alveoli—exchange of gases between air and blood

Lungs and pleura

A Structure

  1 Size—large enough to fill chest cavity except for middle space occupied by heart and large blood vessels

  2 Apex—narrow upper part of each lung, under collarbone

  3 Base—broad lower part of each lung; rests on diaphragm

  4 Pleura—moist, smooth, slippery membrane that lines chest cavity and covers outer surface of lungs; prevents friction between lungs and chest wall during breathing

  5 Respiratory membrane—see Fig. 13-2

B Function—breathing (pulmonary ventilation and pulmonary respiration)

Respiration

A Exchange of gases in lungs

  1 Carbaminohemoglobin breaks down into carbon dioxide and hemoglobin

  2 Carbon dioxide moves out of lung capillary blood into alveolar air and out of body in expired air

  3 Oxygen moves from alveoli into lung capillaries

  4 Hemoglobin combines with oxygen, producing oxyhemoglobin

B Exchange of gases in tissues

  1 Oxyhemoglobin breaks down into oxygen and hemoglobin

  2 Oxygen moves out of tissue capillary blood into tissue cells

  3 Carbon dioxide moves from tissue cells into tissue capillary blood

  4 Hemoglobin combines with carbon dioxide, forming carbaminohemoglobin

C Mechanics of breathing

  1 Contraction of diaphragm and of chest-elevating muscles enlarges chest cavity, expands lungs, and causes air to move down into lungs

  2 Relaxation of diaphragm and of chest elevators decreases size of chest cavity, deflates lungs, and causes air to move out of lungs

D Volumes of air exchanged in pulmonary ventilation—see Fig. 13-7

  1 Volumes of air exchanged in breathing can be measured with a spirometer

  2 Tidal volume (TV) amount normally breathed in and out with each breath

  3 Vital capacity (VC)—largest amount of air that one can breathe in and out in one inspiration and expiration

  4 Expiratory reserve volume (ERV)—amount

of air that can be forcibly inhaled after expiring the tidal volume

5 Inspiratory reserve volume (IRV)—amount of air that can be forcibly inhaled after a normal inspiration

6 Residual volume (RV)—air that remains in the lungs after the most forceful expiration

7 Rate—usually about 16 to 20 breaths a minute; much faster during exercise

Disorders of respiratory system

A Tuberculosis
B Lung cancer
C Pneumonia
D Brain hemorrhage
E Emphysema

New words

adenoids
carbaminohemoglobin
conchae
emphysema
expiratory reserve
  volume (ERV)
hyperventilation
hypoxia
inspiratory reserve
  volume (IRV)
oxyhemoglobin

pleurisy
pneumothorax
residual volume (RV)
respiration
respiratory membrane
spirometer
surfactant
tidal volume (TV)
thoracentesis
vital capacity (VC)

Review questions

1 Identify (list) the paranasal air sinuses.
2 Discuss the functions of the nose in respiration.
3 What and where are the pharynx and larynx?
4 What structures open into the pharynx?
5 What are the anatomical subdivisions of the pharynx?
6 Where are the tonsils and adenoids located?
7 Why does sinusitis or middle ear infection occur so frequently after a common cold?
8 What function do the C-shaped rings of cartilage serve in the trachea?
9 What and where is the "Adam's apple"?
10 Define the following terms:
parietal pleura   pneumothorax
pleural space   visceral pleura
pleurisy
11 Do breathing and respiration mean the same thing? Define each term.
12 Briefly explain how O_2 and CO_2 can move between alveolar air, blood, and tissue cells.
13 Explain the following equation:

$$VC = TV + IRV + ERV$$

14 What does the term "residual volume" mean?
15 How can a brain hemorrhage affect the respiratory system?
16 Discuss the anatomy of the "respiratory membrane."

chapter 14

The urinary system

The urinary system, as you might guess from its name, performs the functions of secreting urine and eliminating it from the body. What you might not guess so easily is how essential these functions are for healthy survival. Unless the urinary system operates normally, the normal composition of blood cannot long be maintained, and serious consequences soon follow. In this chapter we shall discuss the structure and function of each of the urinary system's organs. We shall also mention briefly some disease conditions produced by abnormal functioning of the urinary system.

Kidneys
Location

Almost everyone has two kidneys. To locate them on your own body, stand erect and put your hands on your hips with your thumbs meeting over your backbone. In this position your kidneys lie just above your thumb. Usually the right kidney is a little lower than the left. They are located under the muscles of the back and behind the parietal peritoneum (the membrane that lines the abdominal cavity). This is a convenient anatomical fact. Because of it a surgeon can operate on a kidney without cutting through the peritoneum. A heavy cushion of fat normally encases each kidney and helps hold it in place. In an extremely thin person who lacks this cushion, one or both kidneys may drop down (renal ptosis) enough to put a kink in the tubes that drain urine out of the kidney. This, of course, obstructs urine flow (Fig. 14-1).

Internal structure

If you were to slice through a kidney from side to side and open it like the pages of a book, you would see the structures shown in Fig. 14-2. Identify each of the following:

1. *Cortex*—the outer part of the kidney (The word "cortex" comes from the Latin word for bark or rind, so the cortex of an organ is its outer layer. Kidneys, brain, and adrenal glands all have a cortex.)
2. *Medulla*—the inner portion of the kidney
3. *Pyramids*—triangular-shaped divisions of the medulla of the kidney
4. *Papilla* (pl. *papillae*)—narrow, innermost end of a pyramid
5. *Pelvis*—the kidney or renal pelvis is an expansion of the upper end of a ureter (the tube that drains urine into the bladder)
6. *Calyx* (pl. *calices*)—each calyx is a division of the renal pelvis; opening into each calyx is the papilla of a pyramid

Microscopic structure

More than a million microscopic-sized units named *nephrons* make up each kidney's interior. The shape of a nephron is unique, unmistakable, and admirably suited to its function of producing urine. It looks a little like a tiny funnel with a very long stem, but an unusual stem in that it is highly convoluted, that is, has many bends in it. Locate each of the following parts of a nephron in Fig. 14-3:

1. *Bowman's capsule*—the cup-shaped top of a nephron
2. *Glomerulus*—a network of blood capillaries tucked into the Bowman's capsule
3. *Renal corpuscle*—a Bowman's capsule together with its glomerulus
4. *Proximal convoluted tubule*—the first segment of a renal tubule, called proximal because it lies nearest the tubule's origin from Bowman's capsule and convoluted because it has several bends in it
5. *Loop of Henle*—the extension of the proximal tubule; observe that the loop of Henle consists of a straight descending limb, a loop, and a straight ascending limb
6. *Distal convoluted tubule*—the part of the tubule distal to the ascending limb of Henle, the extension of the ascending limb
7. *Collecting tubule*—a straight, that is, not convoluted, part of a renal tubule; distal tubules of several nephrons join to form a single collecting tubule

In summary, nephrons, the microscopic units of a kidney, have two main parts, a renal corpuscle (Bowman's capsule with glomerulus) and a renal tubule.

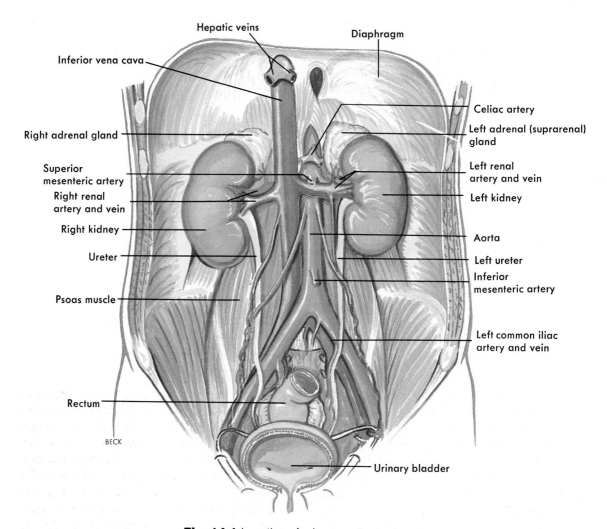

Fig. 14-1 Location of urinary system organs.

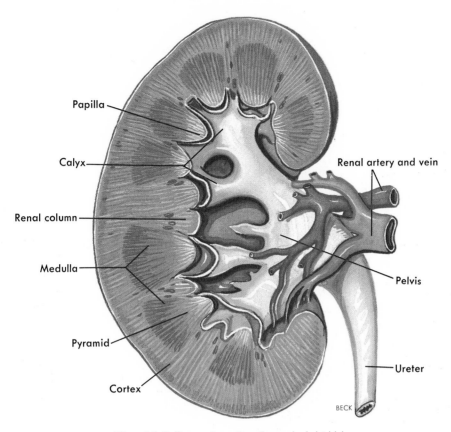

Papilla

Calyx

Renal column

Medulla

Pyramid

Cortex

Renal artery and vein

Pelvis

Ureter

BECK

Fig. 14-2 Coronal section through right kidney.

Function

The kidneys are vital organs. The function they perform, that of forming urine, is essential for maintaining life. In the process of forming urine, fluid, electrolytes, and wastes from metabolism are excreted, that is, they leave the blood and enter the urine. Normally the kidneys vary the amounts of these substances leaving the blood so that they equal the amounts of them entering the blood from various sources. In short, they adjust their output to equal their intake. By so doing, the kidneys play an essential part in maintaining homeostasis. Homeostasis cannot be maintained—nor can life itself—if the kidneys fail and the condition is not soon corrected.

How kidneys form urine

The kidneys' 2 million or more nephrons form urine by three processes: filtration, reabsorption, and secretion. Urine formation begins with the process of *filtration*, which goes on continually in the renal corpuscles (glomeruli plus Bowman's capsules encasing them). Blood flowing through the glomeruli exerts pressure, and this glomerular blood pressure is sufficiently high to push water and dissolved substances out of the glomeruli into the Bowman's capsules. Briefly, glomerular blood pressure causes filtration through the glomerular-capsular membrane. If glomerular blood pressure drops below a certain level, filtration and urine formation cease. Hemorrhage, for ex-

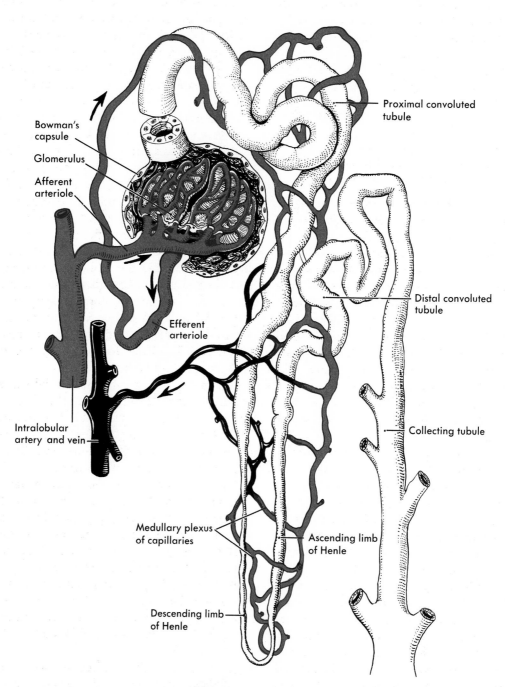

Bowman's capsule

Glomerulus

Afferent arteriole

Efferent arteriole

Intralobular artery and vein

Proximal convoluted tubule

Distal convoluted tubule

Collecting tubule

Medullary plexus of capillaries

Ascending limb of Henle

Descending limb of Henle

Fig. 14-3 A nephron and its blood vessels. Each kidney contains about 1 million nephrons. Arrows indicate the direction of blood flow: from the intralobular artery → afferent arteriole → glomerulus → efferent arteriole → peritubular capillaries (around the tubules) → venules (shown in black) → intralobular vein.

ample, may cause a precipitous drop in blood pressure followed by kidney failure.

Glomerular filtration normally occurs at the rate of 125 milliliters (ml) per minute. The following simple calculations may help you visualize how enormous this volume is:

125 × 60 = 7,500 ml = Glomerular filtration
rate per hour

7,500 × 24 = 180,000 ml or 180 liters (about
190 quarts) = Volume of fluid filtered out of
glomerular blood per day

Obviously no one ever excretes anywhere near 180 liters of urine per day. Why? Because most of the fluid that leaves the blood by glomerular filtration, the first process in urine formation, returns to the blood by the second process—reabsorption.

Reabsorption, by definition, is the movement of substances out of the renal tubules into the blood capillaries located around the tubules (peritubular capillaries). Substances reabsorbed are water, glucose and other nutrients, and sodium and other ions. Reabsorption begins in the proximal convoluted tubules and continues in the loop of Henle, distal convoluted tubules, and collecting tubules.

Large amounts of water—approximately 178 liters per day—are reabsorbed by osmosis from the proximal tubules. In other words, nearly 99% of the 180 liters of water that leave the blood each day by glomerular filtration returns to the blood by proximal tubule reabsorption.

The nutrient glucose is entirely reabsorbed from the proximal tubules. It is actively transported out of them into peritubular capillary blood. None of this valuable nutrient is wasted by being lost in the urine. However, exceptions to this normal rule do occur. For example, in diabetes mellitus, if blood glucose concentration increases above a certain level, the tubular filtrate then contains more glucose than

kidney tubule cells can reabsorb. Some of the glucose therefore remains behind in the urine. Glucose in the urine (glycosuria or glucosuria) is a well-known sign of diabetes.

Sodium ions and other ions are only partially reabsorbed from renal tubules. For the most part sodium ions are actively transported back into blood from the tubular urine. The amount of sodium reabsorbed varies from time to time; it depends largely on salt intake. In general the greater the amount of salt intake, the less the amount of salt reabsorption and therefore the greater the amount of salt excreted in the urine. Also, the less the salt intake, the greater the salt reabsorption and the less salt excreted in the urine. By varying the amount of salt reabsorbed, the body usually can maintain homeostasis of the blood's salt concentration. This is an extremely important matter because cells are damaged by either too much or too little salt in the fluid around them.

Secretion is the process by which substances move into urine in the distal and collecting tubules from blood in the capillaries around these tubules. In this respect secretion is reabsorption in reverse. Whereas reabsorption moves substances out of the urine into the blood, secretion moves substances out of the blood into the urine. Substances secreted are hydrogen ions, potassium ions, ammonia, and certain drugs. Hydrogen ions, potassium ions, and drugs are secreted by being actively transported out of blood into tubular urine. Ammonia is secreted by diffusion. Kidney tubule secretion plays a crucial role in maintaining the body's acid-base balance (see Chapter 18).

In summary, three processes occurring in successive portions of the nephron accomplish the function of urine formation (Fig. 14-4 and Table 14-1):

1. *Filtration*—of water and dissolved sub-

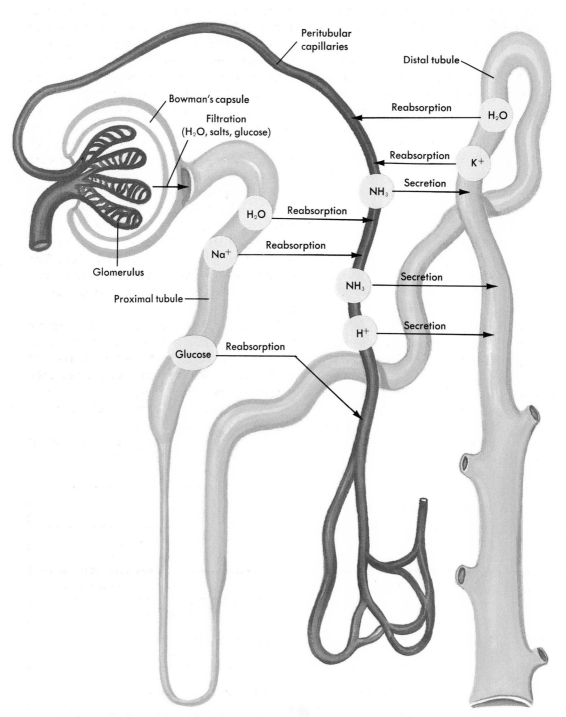

Fig. 14-4 Diagram showing the steps in urine formation in successive parts of a nephron: filtration, reabsorption, and secretion.

stances out of the blood in the glomeruli into Bowman's capsules

2. *Reabsorption*—of water and dissolved substances out of the kidney tubules back into the blood (Note that this process prevents substances needed by the body from being lost in the urine. Usually 97% to 99% of the water filtered out of the glomerular blood is retrieved from the tubules.)

3. *Secretion*—of hydrogen ions (H^+), potassium ions (K^+), and certain drugs

Control of urine volume

The body has ways to control both the amount and the composition of the urine that it secretes. It does this mainly by controlling the amount of water and dissolved substances reabsorbed by the convoluted tubules. For example, a hormone (antidiuretic hormone or ADH) from the posterior pituitary gland tends to decrease the amount of urine by making distal and collecting tubules permeable to water. If no ADH is present, both distal and collecting tubules are practically impermeable to water, so little or no water is reabsorbed from

them. When ADH is present in the blood, distal and collecting tubules are permeable to water and water is reabsorbed from them. As a result, less water is lost from the body or more water is retained—whichever way you wish to say it. At any rate, for this reason ADH is accurately described as the "water-retaining hormone." You might also think of it as the "urine-decreasing hormone." The hormone aldosterone, secreted by the adrenal cortex, plays an important part in controlling the kidney tubules' reabsorption of salt. Primarily it stimulates the tubules to reabsorb sodium salts at a faster rate. Secondarily, aldosterone tends also to increase tubular water reabsorption. The term "salt- and water-retaining hormone" therefore is a descriptive nickname for aldosterone.

Sometimes the kidneys do not excrete normal amounts of urine—as a result of kidney disease, cardiovascular disease, or stress. Here are some terms associated with abnormal amounts of urine:

1. *Anuria*—literally, absence of urine
2. *Oliguria*—scanty urine
3. *Polyuria*—usually large amounts of urine

Table 14-1 Functions of parts of nephron in urine formation

Part of nephron	Process in urine formation	Substances moved and direction of movement
Glomerulus	Filtration	Water and solutes (for example, sodium and other ions, glucose and other nutrients) filter out of glomeruli into Bowman's capsules
Proximal tubule	Reabsorption	Water and solutes
Loop of Henle	Reabsorption	Sodium and chloride ions
Distal and collecting tubules	Reabsorption	Water, sodium, and chloride ions
	Secretion	Ammonia, potassium ions, hydrogen ions, and some drugs

Ureters

Urine drains out of the collecting tubules of each kidney into the renal pelvis and on down the ureter into the urinary bladder (Fig. 14-1). The *renal pelvis* is the basinlike upper end of the ureter and is located inside the kidney. Ureters are narrow tubes less than ¼ inch wide but 10 to 12 inches long. Mucous membrane lines both ureters and renal pelves. Smooth muscle fibers in the ureter walls contract to produce a peristaltic movement that pushes urine down the ureters into the bladder.

Urinary bladder

Elastic fibers and involuntary muscle fibers in the wall of the urinary bladder make it well suited for its functions of expanding to hold variable amounts of urine and then contracting to empty itself. Most people feel the desire to void when the bladder contains about ½ pint (250 milliliters) of urine. Mucous membrane lines the urinary bladder. It lies in folds (rugae) when the bladder is empty.

Urinary *retention* is a condition in which no urine is voided. The kidneys secrete urine, but the bladder for one reason or another cannot empty itself. In urinary *suppression* the opposite is true. The kidneys do not secrete any urine, but the bladder retains the ability to empty itself. Do you think anuria could be a symptom of either suppression or retention?

Incontinence is a condition in which the patient voids urine involuntarily. It frequently occurs in patients who have suffered a stroke or spinal cord injury.

Urethra

To leave the body, urine passes from the bladder, down the urethra, and out its external opening, the *urinary meatus*. In other words, the urethra is the lowermost part of the urinary tract. The same sheet of mucous membrane that lines the renal pelves, ureters, and bladder extends down into the urethra too—a structural feature worth noting because it accounts for the fact that an infection of the urethra may spread upward throughout the urinary tract. The urethra is a narrow tube. It is only about 1½ inches long in a woman but about 8 inches long in a man.

Outline summary

Kidneys

A Location—under back muscles, behind parietal peritoneum, just above waistline; right kidney usually a little lower than left

B Internal structure
   1 Cortex—outer layer of kidney substance
   2 Medulla—inner portion of kidney
   3 Pyramids—triangular-shaped divisions of medulla
   4 Papillae—narrow, innermost ends of pyramids
   5 Pelvis—expansion of upper end of ureter
   6 Calyces—divisions of renal pelvis

C Microscopic structure—nephrons are microscopic units of kidneys; consist of following parts
   1 Bowman's capsule—cup-shaped top of nephron
   2 Glomerulus—network of blood capillaries tucked into the Bowman's capsule
   3 Renal corpuscle—Bowman's capsule with its glomerulus
   4 Proximal convoluted tubule—first segment of a renal tubule
   5 Loop of Henle—extension of proximal tubule; consists of descending limb, loop, and ascending limb
   6 Distal convoluted tubule—extension of ascending limb of Henle
   7 Collecting tubule—straight extension of distal tubule

D Function—urine formation

E How kidneys form urine—by three processes that take place in successive parts of nephron
   1 Filtration—goes on continually in renal cor-

puscles; glomerular blood pressure causes water and dissolved substances to filter out of glomeruli into Bowman's capsule; normal glomerular filtration rate 125 milliliters per minute

   2 Reabsorption—movement of substances out of renal tubules into blood in peritubular capillaries; substances reabsorbed are water, nutrients, and various ions; water reabsorbed by osmosis from proximal tubules

   3 Secretion—movement of substances into urine in the distal and collecting tubules from blood in peritubular capillaries; hydrogen ions, potassium ions, and certain drugs are secreted by active transport; ammonia is secreted by diffusion

F Control of urine volume—mainly by posterior pituitary hormone ADH, which acts to decrease urine volume

Ureters

A Structure—narrow long tubes with expanded upper end (renal pelvis) located inside kidney and lined with mucous membrane

B Function—drain urine from renal pelvis to urinary bladder

Urinary bladder

A Structure
   1 Elastic muscular organ, capable of great expansion
   2 Lined with mucous membrane arranged in rugae, like stomach mucosa

B Functions
   1 Stores urine before voiding
   2 Voiding

Urethra

A Structure
   1 Narrow short tube from urinary bladder to exterior
   2 Lined with mucous membrane
   3 Opening of urethra to exterior called urinary meatus

B Functions
   1 Serves as passageway by which urine leaves bladder for exterior
   2 Passageway by which male reproductive fluid leaves body

New words

anuria	polyuria
Bowman's capsule	renal ptosis
glycosuria	urinary retention
oliguria	urinary suppression

Review questions

  1 What organs form the urinary system?
  2 To operate on a kidney, does a surgeon have to cut through the peritoneum? Explain your answer.
  3 Which kidney usually lies a little lower than the other?
  4 Name the parts of a nephron.
  5 What and where are the glomeruli and Bowman's capsules?
  6 Explain briefly the functions of the glomeruli and Bowman's capsules.
  7 Explain briefly the function of the renal tubules.
  8 What kind of membrane lines the urinary tract?
  9 Explain briefly the function of ADH. What is the full name of this hormone? What gland secretes it?
10 Suppose that ADH secretion increases noticeably. Would this increase or decrease urine volume? Why?
11 What hormone might appropriately be nicknamed the "water-retaining hormone"?
12 What hormone might appropriately be nicknamed the "salt- and water-retaining hormone"?
13 What hormone might be called the urine-decreasing hormone?
14 What is the urinary meatus?
15 What and where are the ureters and the urethra?

Systems that reproduce the body

chapter 15

The male reproductive system

We truly are "fearfully and wonderfully made." Almost any one of the body's structures or functions might have inspired this statement, but of them all perhaps the reproductive systems best deserve such praise. Their achievement is the miracle of duplicating the human body. Their goal is the survival of the human species. Other body systems are concerned primarily with survival of the individual. The reproductive systems are important not only to the individual but to the entire human race.

As you probably noticed, the plural "reproductive systems" was used in the preceding paragraph. The male reproductive system consists of one group of organs and the female reproductive system consists of another group. These two systems differ in structure, but they share a common function—that of reproducing the human body. This chapter discusses the male reproductive system and Chapter 16 discusses the female reproductive system.

Structural plan

So many organs make up the male reproductive system that we need to look first at the structural plan of the system as a whole. Reproductive organs can be classified as either *essential* or *accessory*. The essential organs of reproduction in both males and females are called the *gonads*. The word "gonad" comes from the Greek word *gonos,* meaning "seed." The gonads of the male

Table 15-1 Male reproductive organs	
Essential organs of reproduction or main sex glands (gonads)	**Accessory organs of reproduction**
Testes (right testis and left testis)	Ducts: epididymis, vas deferens, ejaculatory duct (two each of the preceding), and urethra
	Supportive sex glands: seminal vesicle, bulbourethral or Cowper's gland (two each of the preceding), and prostate gland
	External genitals: scrotum and penis

consist of a pair of main sex glands called the *testes.* The seeds produced are the male sex cells or *spermatozoa.*

The accessory organs of reproduction in the male consist of a series of ducts that carry the male sex cells from the testes to the exterior, additional sex glands and the external reproductive organs. Table 15-1 lists the names of all of these structures, and Fig. 15-1 shows the location of most of them.

External genitals

Male external reproductive organs (genitals) consist of two organs, the scrotum and the penis. The scrotum is a skin-covered pouch suspended from the groin region. Internally, it is divided into two sacs by a septum; each sac contains a testis, epididymis, and the lower part of the vas deferens

(Fig. 15-1). The penis is made up of three separate structures: one corpus cavernosum urethrae and two corpora cavernosa penis. These are long Latin names, but not too difficult to remember if you know what they mean. Corpus means "body"; cavernosum means "full of small cavities"—a good description of the erectile tissue that composes all three of these cylindrical bodies. Both "urethrae" and "penis" are in the Latin possessive case and mean "of the urethra" and "of the penis."

Skin covers both the scrotum and the penis. At the distal end of the penis, there is a slightly bulging structure, the *glans penis,* over which the skin is folded doubly to form a more or less loose-fitting, retractable casing called the foreskin (or prepuce). If the foreskin fits too tightly about the glans, a *circumcision* (surgical removal of the foreskin) is usually performed to prevent irritation.

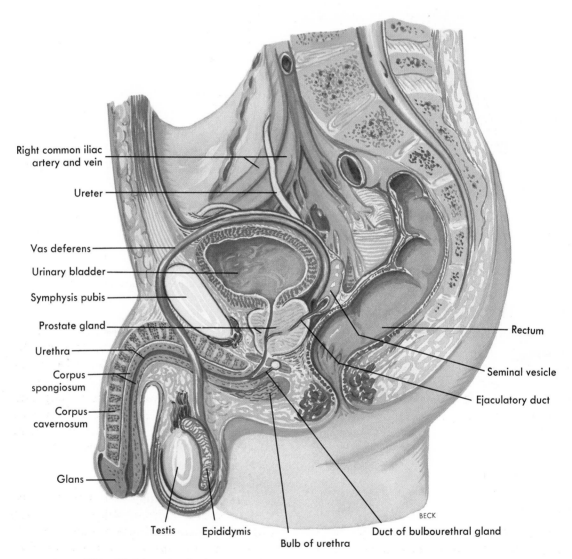

Right common iliac
artery and vein

Ureter

Vas deferens

Urinary bladder

Symphysis pubis

Prostate gland

Urethra

Corpus
spongiosum

Corpus
cavernosum

Glans

Testis   Epididymis

Bulb of urethra

Rectum

Seminal vesicle

Ejaculatory duct

Duct of bulbourethral gland

BECK

Fig. 15-1 Longitudinal section of the male pelvis showing the location of the male reproductive organs.

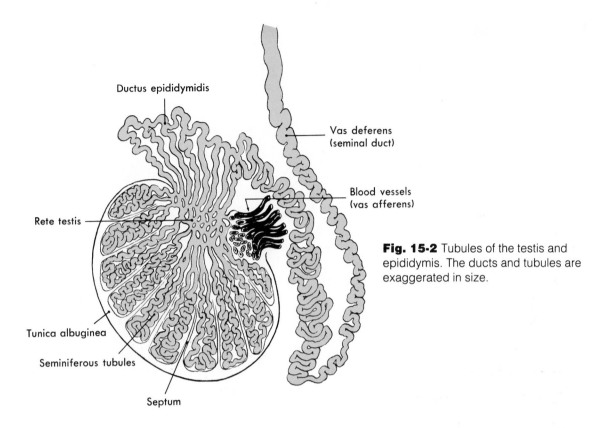

Ductus epididymidis

Vas deferens
(seminal duct)

Blood vessels
(vas afferens)

Rete testis

Tunica albuginea

Seminiferous tubules

Septum

Fig. 15-2 Tubules of the testis and epididymis. The ducts and tubules are exaggerated in size.

The penis is the male organ of coitus (sexual intercourse). Under the stimulus of sexual emotion, blood floods the spaces in the erectile tissue of the three corpora, distending them enough to produce erection of the penis.

Testes

Structure

The testes are the gonads of the male. They are small oval-shaped glands about 1½ inches (3.8 centimeters) long and 1 inch (2.5 centimeters) wide. They are shaped somewhat like an egg that has been flattened slightly from side to side. Note in Fig. 15-2 that each testis is surrounded by a tough whitish membrane called the *tunica albuginea*. This membrane covers the testi-

cle and then enters the gland to form the many septa that divide it into sections or *lobules*. As you can see in Fig. 15-2, each lobule consists of a narrow but long and coiled *seminiferous tubule*. Clusters of cells lie between the seminiferous tubules. These are the *interstitial cells* of the testes.

Functions

To judge the testes by their size would be to underestimate their importance. Each testis forms many millions of male sex cells *(spermatozoa* or *sperm)* every month after puberty. Any one of these tiny cells may join with a female sex cell *(ovum)* to become a new human being (Fig. 15-3). Each testis also secretes the male sex hormone testosterone, the hormone that in a few short months transforms a boy into a man. Testosterone lowers the pitch of his voice,

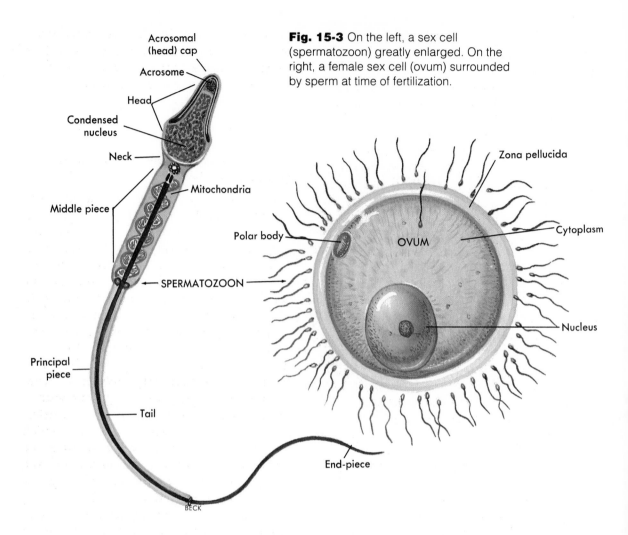

Fig. 15-3 On the left, a sex cell (spermatozoon) greatly enlarged. On the right, a female sex cell (ovum) surrounded by sperm at time of fertilization.

makes his muscles grow large and strong, and even changes the size and shape of his bones.

From the age of puberty on, the seminiferous tubules are almost continuously forming spermatozoa or sperm. *Spermatogenesis* is the name of this process. The other function of the testes is to secrete the male hormone testosterone. This function, however, is carried on by the *interstitial cells* of the testes, not by its seminiferous tubules.

The testis, then, is a beautiful example of the familiar principle that "structure determines function" (specifically, the production of sperm by the seminiferous tu-

bules and secretion of testosterone by the interstitial cells).

Testosterone

Testosterone serves the following general functions:

1 It masculinizes. The various characteristics that we think of as "male" develop because of testosterone's influence. For instance, when a young boy's voice changes, it is testosterone that brings this about.

2 It promotes and maintains the development of the male accessory organs (prostate gland, seminal vesicles, and so on).

3 It has a stimulating effect on protein

anabolism. Testosterone thus is responsible for the greater muscular development and strength of the male.

A good way to remember testosterone's functions is to think of it as "the masculinizing hormone" and "the anabolic hormone."

How much testosterone the interstitial cells of the testes secrete depends on how much of another hormone—the interstitial cell–stimulating hormone (ICSH)—the anterior pituitary gland secretes. The more ICSH screted, the more it stimulates the interstitial cells to secrete more testosterone. In short, a high blood concentration of ICSH stimulates testerone secretion. The resulting high blood concentration of testosterone then feeds back via the circulating blood to influence ICSH secretion by the anterior pituitary gland, but here the effect is quite different. Testosterone exerts a negative effect on ICSH secretion. A high blood concentration of testosterone inhibits ICSH secretion instead of stimulating it. This is an example of a "negative feedback control mechanism"—a term used frequently in our computer-conscious world.

Spermatozoa

Spermatozoa are among the smallest and most highly specialized cells in the body. Ejaculation of sperm into the female vagina during sexual intercourse is only one step in the long journey that these sex cells must make before they can meet and fertilize an ovum. Sperm cells are, in effect, specialized packages of genetic information. All of the characteristics that a baby will inherit from its father at fertilization are contained in the condensed nuclear (genetic) material found in each sperm head. All other parts of the sperm cell are designed to provide motility. Fertilization normally occurs in the fallopian tubes or oviducts of the female. Therefore, in order for fertilization to occur, sperm must "swim" for relatively long

distances through the female reproductive ducts. Fig. 15-3 shows the characteristic parts of a single spermatozoon: head, neck, middle piece, and elongated, lashlike tail.

Ducts
Epididymis, vas deferens, ejaculatory duct, and urethra

The role of the vas deferens in birth control has recently become very important. To find out why, we need first to trace the route by which sperm leave the male reproductive tract in order to enter the female tract. As you read the description of this route in the next sentences, follow it in Figs. 15-1 and 15-2. Sperm are formed in the testes by the seminiferous tubules. From the seminiferous tubules, sperm by the millions stream into a narrow but long and tightly coiled duct, the *epididymis*. From the duct of the epididymis sperm continue on their way through one of the vasa deferentia into an ejaculatory duct from which they move down the urethra and out of the body. Note in Fig. 15-1 the location of the vas deferens in the scrotum. There are a pair of these small tubes and here, in the scrotum, they lie near the surface. This fact makes it possible for a surgeon to make a small incision in the scrotum and quickly and easily cut out a section of each vas and tie off each of its separated ends. The technical name for this minor surgery is a bilateral partial *vasectomy*. Although a man's seminiferous tubules may continue to form sperm after he has had a vasectomy, they can no longer leave his body. A part of their exit route has been cut away and no detour route has been provided. Therefore the man has become sterile, that is, he can no longer father children. If at a later date the man changes his mind and wishes to be fertile again, it is technically possible in some cases to rejoin the separated ends of the vas deferens. Unfortunately, however, this does not guarantee that the man's fertility will be restored. Like marriage, therefore, a vasectomy "should not be entered into lightly." A vasectomy does not change a man's ability to have an erection or an ejaculation. As explained in the next paragraph, structures that lie beyond the vas deferens produce the semen.

Accessory male reproductive glands

The two seminal vesicles, one prostate gland, and two bulbourethral (Cowper's) glands are accessory male glands that produce alkaline secretions. These secretions constitute the gelatinous fluid part of the *semen*. Usually 3 to 5 milliliters (about 1 teaspoonful) of semen is ejaculated at one time, and each milliliter contains over 60 million sperm. After a successful bilateral vasectomy, about the same amount of fluid may be ejaculated as before but it contains no sperm. In short, a vasectomy makes a man sterile but not impotent.

The *prostate gland* claims importance not so much for its function as for its trouble-making. In older men it often becomes inflamed and enlarged, squeezing on the urethra, which runs through the center of the doughnut-shaped prostate. Sometimes, in fact, the prostate enlarges so much that it closes off the urethra completely. Urination then becomes impossible. (Should you refer to this as urinary retention or urinary suppression? If you are not sure, check your answer on p. 212.)

The small *bulbourethral* (Cowper's) *glands* lie one on either side of the urethra just below the prostate gland. Like the seminal vesicles and the prostate, the bulbourethral glands add an alkaline secretion to the semen. Sperm survive and remain fertile longer in an alkaline fluid than in an acid one.

Outline summary
Structural plan

Reproductive organs are classified as either *essential* or *accessory*. The essential organs (gonads) in the male are the testes. The accessory organs of reproduction consist of a series of ducts that carry the male sex cells from the testes to the exterior, additional sex glands and the external organs (genitals)—see Table 15-1

External genitals

A Scrotum—a skin-covered pouch; each half of scrotum containing a testis, epididymis, and lower part of a vas deferens
B Penis—composed of one corpus cavernosum urethrae and two corpora cavernosa penis; erectile tissue makes up corpora; glans penis, bulging distal end of the organ; foreskin, a double fold of skin, covers glans in uncircumcised male; penis functions as organ of coitus; erection produced by blood distending spaces in erectile tissue

Testes

A Structure—pair of small oval glands in the scrotum; tough membrane encloses each testis and forms partitions (septa) inside, dividing it into lobules
B Functions—serve as essential male gland (gonad); seminiferous tubules form male sex cells (process of spermatogenesis); interstitial cells secrete male hormone (testosterone)
C Testosterone
  1 Functions—masculinizes, promotes and maintains development of accessory male organs, stimulates protein anabolism
  2 Control of secretion—high blood level of ICSH stimulates testosterone secretion; conversely, high blood level of testosterone inhibits ICSH secretion—example of "negative feedback control mechanism"
D Spermatozoa—Fig. 15-3 shows the characteristic parts of a single spermatozoon: head, neck, middle piece, and elongated, lashlike tail

Ducts

A Epididymis—narrow tube attached to each testis; duct of testis

B Vas deferens—continuation of ducts that start in epididymis

C Ejaculatory duct—continuation of vas deferens

D Urethra—terminal duct in male of both reproductive and urinary tracts

Accessory male reproductive glands

A Seminal vesicles—secrete alkaline substance into semen

B Prostate gland—encircles urethra just below bladder; secretes alkaline fluid into semen

C Bulbourethral glands—pair of small glands located just below prostate; ducts open into urethra where they add alkaline secretion to semen

Review questions

1 Identify the essential and accessory organs of reproduction in the male.

2 What organs are included in the "external genitals" of the male?

3 Discuss the structure of the testes. Explain the function of the seminiferous tubules and the interstitial cells of the testes.

4 What is the name of the masculinizing hormone? What glands secrete it? What are its general functions? How is its secretion level controlled?

5 Discuss the anatomy of a sperm cell. Why is it motile? What parts of the sperm cell are designed to provide motility?

6 What structures form semen?

7 Trace a sperm cell from its point of formation in the testes through the male reproductive ducts to ejaculation.

8 Explain why a vasectomy sterilizes a male.

9 How many sperm are normally present in one ejaculation of semen?

10 What is the usual volume of semen ejaculated at one time?

11 Castration is an operation that removes the testes. Would this sterilize the male? Why? What other effects would you expect to see as a result of castration? Why?

chapter 16

The female reproductive system

Structural plan

The structural plan of the female reproductive system resembles that of the male system. Like the male reproductive system, the female reproductive system consists of *essential* and *accessory* organs of reproduction. The essential organs of reproduction in both sexes are called *gonads*. The gonads of the female consist of a pair of main sex glands called *ovaries*. The female sex cells or *ova* are produced here.

The accessory organs of reproduction in the female consist of ducts from the main sex glands to the exterior, additional sex glands, and the external reproductive organs (genitals). To find out the names of these structures in the female, consult Table 16-1 and Figs. 16-1 and 16-2.

External genitals

The scientific name for the female external genitals is the *vulva (pudenda).* Identify the following structures of the vulva in Fig. 16-2:

1. Mons pubis
2. Clitoris
3. Orifice of urethra
4. Labia minora (small lips)
5. Hymen
6. Orifice, duct of Bartholin's gland
7. Orifice of vagina
8. Labia majora (large lips)

Table 16-1 Female reproductive organs	
Essential organs of reproduction or main sex glands (gonads)	**Accessory organs of reproduction**
Ovaries (right and left ovaries)	Ducts: uterine tubes (two), uterus, vagina
	Accessory sex glands: Bartholin's glands, breasts
	External genitals: vulva (pudenda) (see Fig. 16-2)

Ovaries

Although male and female reproductive systems resemble each other in plan, they differ from one another in details. For example, the testes, the main sex glands of the male, are not located inside a body cavity but lie in an external skin-covered pouch, the scrotum. In the female, however, the main sex glands, the ovaries, lie within the pelvic cavity. Also, there are two differences between male and female reproductive systems that sometimes assume clinical importance. One is that the ovarian ducts in the female open into the abdominal cavity. They do not attach directly to the ovaries. Later we shall explain the importance of this fact (p. 228). The other clinically significant difference is that the male urethra serves as a duct for both the urinary tract and reproductive tract. This is not the case in the female. In a woman's body the urethra serves as a duct for the urinary tract only. A separate tube, the vagina, serves the reproductive tract as its duct to the outside. This difference can be important. For example, if a man contracts gonorrhea, the infection can easily spread from the urethra throughout both his reproductive and urinary tracts. Although the urinary and reproductive systems are separated in the female, vaginal infections can spread thrugh the uterine tubes directly into the abdomen. In the male there is no such direct route by which microorganisms can reach the abdominal cavity from the exterior.

Several thousands sacs, too small to be seen without a microscope, make up the bulk of each ovary. They are called *ovarian follicles*. Within each follicle lies an immature *ovum*, the female sex cell. A mature ovum in its sac is called a *graafian follicle*, in honor of the Dutch anatomist who discovered them some 300 years ago.

Like the testes, the ovaries produce both sex cells and sex hormones. However, the ovaries usually produce only one mature

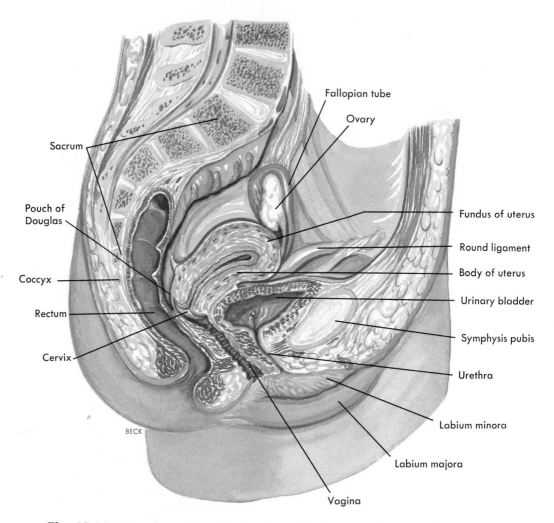

Fig. 16-1 Longitudinal section of the female pelvis showing the location of the female reproductive organs.

sex cell—mature ovum, that is—a month, whereas the testes produce many millions of mature sperm. The ovaries secrete two kinds of female hormones, namely, estrogens and progesterone. The testes, on the other hand, secrete only one kind of male hormone, namely, *androgens*; testosterone is the only important androgen. The only endocrine gland cells of the testes are the interstitial cells. The ovaries contain two endocrine glands, namely, the graafian follicles and the *corpus luteum.* Graafian follicles secrete estrogens. The corpus luteum secretes chiefly progesterone but also some estrogens.

Just as androgens are the masculinizing hormone, so estrogens are the feminizing hormone. To discover estrogens' two chief functions, see Fig. 16-3. Look next at Fig. 16-4; it will tell you the two chief functions progesterone performs.

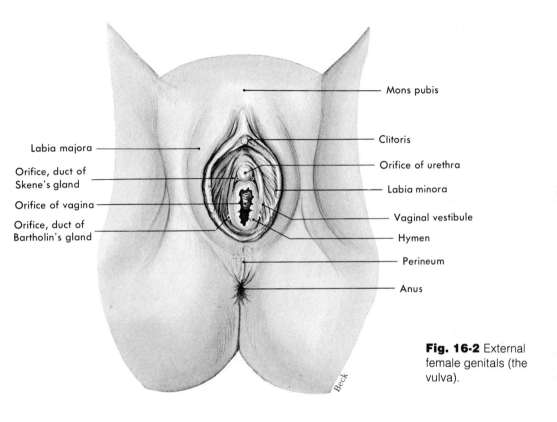

Mons pubis

Clitoris

Orifice of urethra

Labia majora

Orifice, duct of
Skene's gland

Labia minora

Orifice of vagina

Vaginal vestibule

Orifice, duct of
Bartholin's gland

Hymen

Perineum

Anus

Beck

Fig. 16-2 External
female genitals (the
vulva).

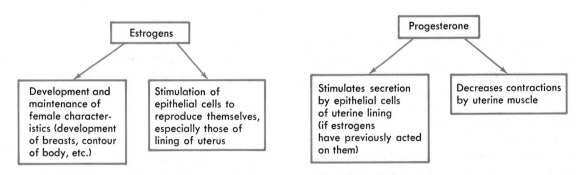

Estrogens

Development and
maintenance of
female character-
istics (development
of breasts, contour
of body, etc.)

Stimulation of
epithelial cells to
reproduce themselves,
especially those of
lining of uterus

Progesterone

Stimulates secretion
by epithelial cells
of uterine lining
(if estrogens
have previously acted
on them)

Decreases contractions
by uterine muscle

Fig. 16-3 Functions of estrogens.

Fig. 16-4 Functions of progesterone.

Ducts

Uterine tubes (fallopian tubes)

The uterine tubes serve as ducts for the ovaries even though they are not attached to them. The outer end of each tube curves over the top of each ovary and opens into the abdominal cavity. The inner end of each uterine tube attaches to the uterus, and the cavity inside the tube opens into the cavity in the uterus. Each tube measures about 4 inches (10 centimeters) in length.

Study Fig. 16-5. After ovulation, the discharged ovum first enters the abdominal cavity and then finds its way into the uterine tube. Fertilization normally occurs in the outer one third of the oviduct. Cells in the fertilized ovum immediately begin to multiply and in about 3 days a solid mass of cells called a *morula* is formed. By the time the developing embryo reaches the uterus it is a hollow ball of cells called a *blastocyst*. Implantation in the uterine lining is soon complete.* Occasionally, however, because the outer ends of the uterine tubes open into the pelvic cavity and are not actually connected to the ovaries, an ovum does not enter a tube but becomes fertilized in the abdominal cavity. The term "ectopic pregnancy" means a pregnancy that develops outside its proper place in the cavity of the uterus.

*On July 25, 1978, the world's first test-tube baby, a girl, was born in Oldham, England. Nine months earlier a mature ovum had been removed from the mother's body by laparoscopy ("belly button surgery") and fertilized in a laboratory dish by her husband's sperm. After 2½ days' growth in a controlled environment, physicians returned the fertilized ovum (now at the eight-cell stage) to the mother's uterus.

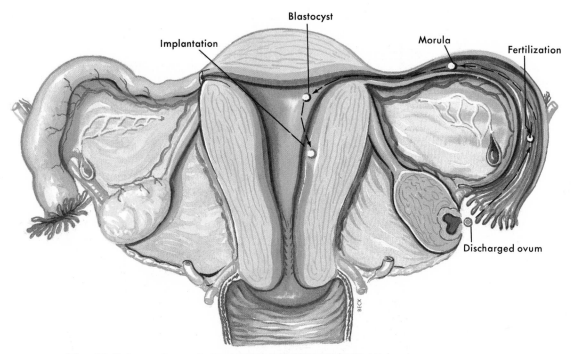

Fig. 16-5 Appearance of uterus and uterine tubes, fertilization to implantation. Note that fertilization normally occurs in the outer one third of the uterine tube.

Uterus

The uterus is a small organ—only about the size of a pear—but it is extremely strong. It is almost all muscle with only a small cavity inside. During pregnancy, the uterus grows many times larger so that it becomes big enough to hold a baby plus a considerable amount of fluid. The uterus is composed of two parts: an upper portion, the *body*, and a lower narrow section, the *cervix*. Just above the level where the uterine tubes attach to the body of the uterus, it rounds out to form a bulging prominence called the *fundus*. (See Fig. 16-6.) Except during pregnancy, the uterus lies in the pelvic cavity just behind the urinary bladder. By the end of pregnancy it becomes large enough to extend up to the top of the abdominal cavity. It then presses against the underside of the diaphragm—a fact that

explains such a comment as "I can't seem to take a deep breath since I've gotten so big," made by many women late in their pregnancies.

The uterus functions in three processes—menstruation, pregnancy, and labor. The corpus luteum stops secreting progesterone and decreases its secretion of estrogens about eleven days after ovulation. Three days later, when the progesterone and estrogen concentrations in the blood are at their lowest, menstruation starts. Small pieces of *endometrium* (mucous membrane lining of the uterus) pull loose, leaving torn blood vessels underneath. Blood and bits of endometrium trickle out of the uterus into the vagina and out of the body. Immediately after menstruation the endometrium starts to repair itself. It again grows thick and becomes lavishly supplied with blood

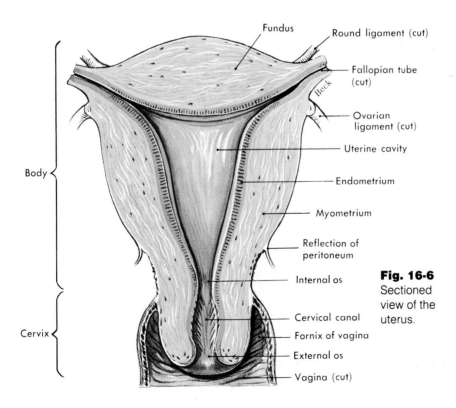

Fundus

Round ligament (cut)

Fallopian tube (cut)

Beck

Ovarian ligament (cut)

Uterine cavity

Endometrium

Myometrium

Reflection of peritoneum

Body

Internal os

Cervix

Cervical canal

Fornix of vagina

External os

Vagina (cut)

Fig. 16-6
Sectioned view of the uterus.

in preparation for pregnancy. If fertilization does not take place, the uterus once more sheds the lining made ready for a pregnancy that did not occur. Because these changes in the uterine lining continue to repeat themselves, they are spoken of as the *menstrual cycle*. For a description of this cycle in the form of a diagram, see Fig. 16-8.

Menstruation first occurs at puberty, often around the age of 12 years. Normally, it repeats itself about every 28 days or thirteen times a year for some 30 to 40 years before it ceases *(menopause* or *climacteric)* when a woman is somewhere around the age of 45 years.

Vagina

The vagina is a distensible tube made mainly of smooth muscle and lined with mucous membrane. It lies in the pelvic cavity between the urinary bladder and the rectum. As the part of the female reproductive tract that opens to the exterior, the vagina is the organ that sperm enter on their journey to meet an ovum, and it is also the organ from which a baby emerges to meet its new world.

Accessory female reproductive glands
Bartholin's glands

One of the small Bartholin's glands lies to the right of the vaginal outlet and one to the left of it. Secreting a mucuslike lubricating fluid is the function of Bartholin's glands. Their ducts open into the space between the labia minora and the hymen (Fig. 16-2). Bartholinitis, an infection of these glands, occurs frequently. It often develops, for example, when a woman contracts gonorrhea.

Breasts

The breasts lie over the pectoral muscles and are attached to them by connective tissue ligaments (of Cooper) (Fig. 16-7). Breast size is determined more by the amount of fat around the glandular (milk-secreting) tissue than by the amount of glandular tissue itself. Hence the size of the breast has little to do with its ability to secrete adequate amounts of milk after the birth of a baby.

Each breast consists of fifteen to twenty divisions or lobes that are arranged radially. Each lobe consists of several lobules, and each lobule consists of milk-secreting (glandular) cells. The milk-secreting cells are arranged in grapelike clusters called *alveoli*. Small *lactiferous ducts* drain the alveoli and converge toward the nipple like the spokes of a wheel. Only one lactiferous duct leads from each lobe to an opening in the nipple. The colored area around the nipple is the *areola*. It changes from a delicate pink to brown early in pregancy and never quite returns to its original color.

A knowledge of the lymphatic drainage of the breast is important because cancerous cells from breast tumors often spread to other areas of the body through the lymphatics. The lymphatic drainage of the breast is discussed in Chapter 10. (See also Fig. 10-15.)

Menstrual cycle
Phases and events

The menstrual cycle consists of a great many changes—in the uterus, ovaries, vagina, and breasts and in the anterior pituitary gland's secretion of hormones. In the majority of women these changes occur with almost precise regularity throughout their reproductive years. The first indication of changes comes with the event of the

Fig. 16-7 Lateral view of the breast (sagittal section). The gland is fixed to the overlying skin and the pectoral muscles by the suspensory ligaments of Cooper. Each lobule of secretory tissue is drained by a lactiferous duct that opens through the nipple.

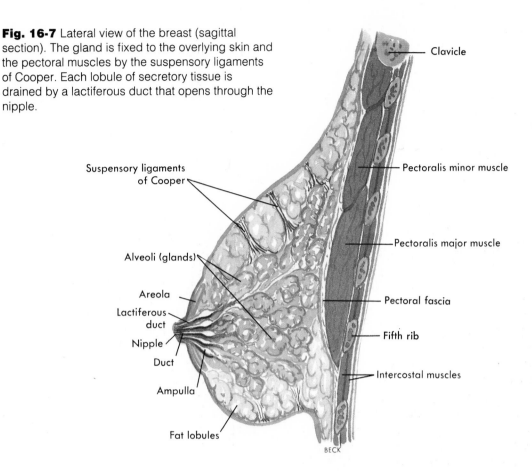

Clavicle

Pectoralis minor muscle

Pectoralis major muscle

Suspensory ligaments of Cooper

Pectoral fascia

Alveoli (glands)

Areola

Fifth rib

Lactiferous duct

Nipple

Duct

Intercostal muscles

Ampulla

Fat lobules

BECK

first menstrual period. (*Menarche* is the scientific name for the beginning of the menses.)

A typical menstrual cycle covers a period of 28 days. Each cycle consists of three phases. Although they have been called by several different names, we shall call them the menstrual period, the postmenstrual phase, and the premenstrual phase. Now examine Fig. 16-9 to find out what changes take place in the lining of the uterus (endometrium) and in the ovaries during each phase of the menstrual cycle. Be sure you do not overlook the event that occurs on day 14 of a 28-day cycle.

As a general rule, only one ovum matures each month during the 30 or 40 years that a woman has menstrual periods. But there are exceptions to this rule. Some months more than one matures, and some months no ovum matures. Ovulation occurs 14 days before the next menstrual period begins. In a 28-day menstrual cycle, this means that ovulation occurs on the fourteenth day of the cycle, as shown in Fig. 16-8. (Note that the first day of the menstrual period is considered the first day of the cycle.) In a 30-day cycle, however, the fourteenth day before the beginning of the next menses is not the fourteenth cycle day, but the sixteenth.

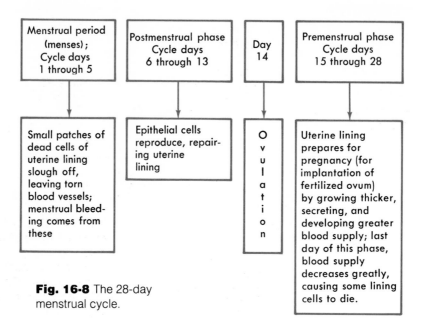

Fig. 16-8 The 28-day menstrual cycle.

And in a 25-day cycle, the fourteenth day before the next menses begins is the eleventh cycle day.

This matter of the time of ovulation has great practical importance. An ovum lives only a short time after it it is ejected from its follicle, and sperm live only a short time after they enter the female body. Fertilization of an ovum by a sperm therefore can occur only around the time of ovulation. In other words, a woman's fertile period lasts only a few days out of each month. Simply knowing the length of previous cycles, however, cannot ensure with any degree of accuracy the length of a current cycle or some future cycle. The reason is that the vast majority of women show some month-to-month variation in the length of their cycles. This physiological fact probably accounts for most of the unreliability of the old "calendar rhythm method" of contraception. Calendar rhythm has now been replaced by other, far more reliable "natural family planning" methods. Such natural methods base their judgments about fertil-

ity on information other than previous cycle lengths, for example, measurement of body temperature and recognition of changes each month in the amount and consistency of cervical mucus. Changes in the blood levels of the hormones that control ovulation also produce changes in body temperature and in the amount and consistency of the cervical mucus.

Knowledge about the body's method of controlling the events of the menstrual cycle has made possible most of our modern methods of birth control. Knowledge about the male reproductive system has led to other methods—vasectomy, for example. Current and proposed research give promise of still other methods that may prove even more practical.

Control of menstrual cycle changes

Now that overpopulation looms so large as a threat to man's healthy survival on this planet, knowledge about the body's methods for controlling menstrual cycle events has taken on new and enormous impor-

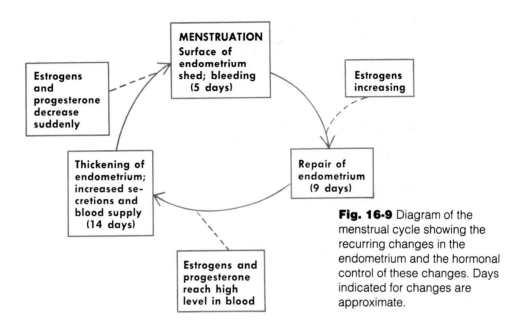

Fig. 16-9 Diagram of the menstrual cycle showing the recurring changes in the endometrium and the hormonal control of these changes. Days indicated for changes are approximate.

tance. Dominating this control is the anterior pituitary gland, the master gland of the body.

From the first to about the seventh day of the menstrual cycle, the anterior pituitary gland secretes increasing amounts of FSH (follicle-stimulating hormone). A high blood concentration of FSH stimulates several immature ovarian follicles to start growing and secrete estrogens (Fig. 16-9). As the estrogen content of blood increases, it stimulates the anterior pituitary gland to secrete another hormone, namely LH (luteinizing hormone). LH causes maturing of a follicle and its ovum, ovulation (rupturing of mature follicle with ejection of ovum), and luteinization (formation of a yellow body, the corpus luteum, in the ruptured follicle).

Which hormone—FSH or LH—would you call the "ovulating hormone"? Do you think ovulation could occur if the blood concentration of FSH remained low throughout the menstrual cycle? If you answered LH to the first question and no to the second, you

answered both questions correctly. Ovulation cannot occur if the blood level of FSH stays low, because a high concentration of this hormone is essential to stimulate ovarian follicles to start growing and maturing. With a low level of FSH, no follicles start to grow, and therefore none become ripe enough to ovulate. Ovulation is caused by the combined actions of FSH and LH. Birth control pills that contain estrogen substances suppress FSH secretion. This indirectly prevents ovulation.

Ovulation occurs, as we have said, because of the combined actions of the two anterior pituitary hormones, FSH and LH. The next question is: what causes menstruation? A brief answer is this: a sudden, sharp decrease in estrogen and progesterone secretion toward the end of the premenstrual period causes the uterine lining to break down and another menstrual period to begin. But exactly what causes the sudden decrease in estrogen and progesterone secretion, we do not yet know.

Disorders of reproductive systems

Infections of the mucous membrane lining of the reproductive tract occur in both sexes. In women the danger of this becomes especially great after childbirth. Physicians and nurses therefore take care not to introduce infectious organisms into the reproductive tract during or after delivery. The name given the inflammation indicates the part of the tract inflamed. For example, vaginitis is inflammation of the vagina; cervicitis, inflammation of the cervix; endometritis, inflammation of the endometrium; salpingitis, inflammation of the uterine tubes. Tumors frequently develop in the uterus, ovaries, and breasts of women and in the prostate gland of men.

Outline summary

Structural plan

Reproductive organs are classified as either *essential* or *accessory*. The essential organs (gonads) in the female are the ovaries. The accessory organs of reproduction consist of ducts from the main sex glands to the exterior, additional sex glands, and the external reproductive organs (genitals)—see Table 16-1

A External genitals, vulva—see Fig. 16-2
B Ovaries—main sex glands of female; located in pelvic cavity; female sex cells (ova) form in graafian follicles in ovaries; graafian follicles secrete estrogens (see Fig. 16-3 for estrogen functions); corpus luteum secretes estrogens and progesterone (see Fig. 16-4 for functions)
C Uterine tubes (fallopian tubes)—ducts for ovaries but not attached to them—see Fig. 16-5

D Uterus (womb)
  1 Parts—fundus, body, cervix—see Fig. 16-6
  2 Structure—strong muscular organ with mucous lining (endometrium)
  3 Functions—menstruation, pregnancy, labor
E Vagina—muscular tube lined with mucous membrane; terminal part of female reproductive tract
F Bartholin's glands—pair of small glands near vaginal orifice; secrete mucuslike lubricating fluid
G Breasts—glands present in both sexes but normally function only in female—see Fig. 16-7

Menstrual cycle

A Length—about 28 days, varies somewhat in different individuals and in the same individual at different times
B Phases
  1 Menstrual period or menses—about the first 4 or 5 days of the cycle, varies somewhat; characterized by sloughing of bits of endometrium (uterine lining) with bleeding
  2 Postmenstrual phase—days between the end of menses and ovulation; varies in length; the shorter the cycle, the shorter the postmenstrual phase; the longer the cycle, the longer the postmenstrual phase; examples, in 28-day cycle postmenstrual phase ends on thirteenth day, but in 26-day cycle it ends on eleventh day, and in 32-day cycle, it ends on seventeenth day; characterized by repair of endometrium
  3 Premenstrual phase—days between ovulation and beginning of next menses; ovulation about 14 days before next menses; characterized by further thickening of endometrium and secretion by its glands in preparation for implantation of fertilized ovum; combined actions of the anterior pi-

tuitary hormones FSH and LH cause ovulation; sudden, sharp decrease in estrogens and progesterone bring on menstruation if pregnancy does not occur

Disorders of reproductive systems

A Infections—common in both sexes; for example, in women, salpingitis, vaginitis, and endometritis, and in men, prostatitis
B Tumors—occur in both sexes, for example, in women, in uterus, ovaries, and breasts, and in men, prostate gland

New words

cervicitis	menses
climacteric	ovulation
endometritis	prostatitis
menarche	vaginitis
menopause	

Review questions

1 Identify the feminizing hormone by its scientific name. What gland secretes it?
2 Identify the ovulating hormone by its scientific name. What glands secrete it?
3 Identify and locate each of the following:

alveoli of breast	graafian follicle
areola	labia majora
cervix	mons pubis
clitoris	ovum
fundus of uterus	uterine tube

4 Define briefly the words listed under "New words."
5 What causes ovulation?
6 What is menstruation?
7 What causes menstruation?
8 How many female sex cells are usually formed each month? How does this compare with the number of male sex cells formed each month?

Fluid, electrolyte, and acid-base balance

chapter 17

Fluid and electrolyte balance

Body fluids

Mechanisms that maintain fluid balance
Importance of electrolytes in body fluids
Capillary blood pressure and blood proteins

Fluid imbalances

Have you ever wondered why sometimes you excrete great volumes of urine and sometimes almost none at all? Why sometimes you feel so thirsty that you can hardly get enough to drink and other times you want no liquids at all? These conditions and many more relate to one of the body's most important functions—that of maintaining its fluid and electrolyte balance. Health and sometimes even survival itself depend on this complex function.

In this chapter you will find a discussion of body fluids and electrolytes, their normal values, the mechanisms that operate to keep them normal, and some of the more common types of fluid and electrolyte imbalances.

Body fluids

If you are a healthy young person and you weigh 120 pounds, there is a good chance that out of the hundreds of compounds present in your body, one substance alone weighs about 72 pounds, or 60% of your total weight! This, the body's most abundant compound, is water. It occupies three main locations known as fluid compartments. Look now at Fig. 17-1. Note that the largest volume of water by far lies inside cells and that it is called, appropriately, *intracellular fluid* (ICF). Note, too, that the water outside of cells—*extracellular fluid* (ECF)—is located in two compartments; in the microscopic spaces between cells where it is called *interstitial fluid* (IF) and in the blood vessels where it is called *plasma*. (Plasma is the liquid part of the blood and it constitutes a little less than half of the total blood volume; blood cells make up the rest of the volume.)

A normal body maintains fluid balance. The term "fluid balance" means that the volumes of ICF, IF, and plasma and the total volume of water in the body all remain relatively constant. Of course not all bodies contain the same amount of water. The more a person weighs, the more water his body contains. This is true because, excluding fat or adipose tissue, about 70% of the body weight is water. Since fat is almost water free, the more fat present in the body, the less the total water content will be per unit of weight. In other words, fat people have a lower water content per pound of body weight than slender people. A slender adult body, for instance, typically consists of about 60% water. An obese body, in contrast, may consist of only 50% or even less water.

Sex and age also influence how much of the body's weight consists of water. Infants have more water in comparison to body weight than adults of either sex. In a newborn infant water may account for 80% of the total body weight. There is a rapid decline in the proportion of body water to body weight during the first year of life. Fig. 17-2 illustrates the proportion of body weight represented by water in newborn infants (80%); adult males (55%); adult females (45%); and adults of either sex without body fat (70%). The female body contains slightly less water per pound of weight because it contains slightly more fat than the male body. Age and the body's water content are inversely related. In general, as age increases, the amount of water per pound of body weight decreases. Inversely related also (as explained in the preceding paragraph) are the amounts of fat and water in the body. In general, the more fat, the less water per pound of weight.

Mechanisms that maintain fluid balance

The body's chief mechanism for maintaining fluid balance is to adjust its fluid output so that it equals its fluid intake. Obviously as long as output and intake are equal, the total amount of water in the body does not change. Fig. 17-3 shows the three sources of fluid intake: the liquids we drink, the water in the foods we eat, and the water formed by catabolism of foods. Fig. 17-3 also indicates that fluid output from the body occurs by way of four organs—the kidneys, lungs, skin, and intestines. The fluid output that changes the most is that from the kidneys. The body maintains fluid balance mainly by changing the volume of urine excreted to match changes in the volume of fluid intake. Everyone knows this from experience. The more liquid one drinks, the more urine one excretes. And conversely, the less the fluid

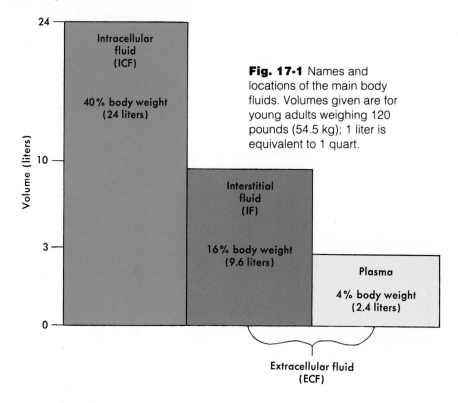

Fig. 17-1 Names and locations of the main body fluids. Volumes given are for young adults weighing 120 pounds (54.5 kg); 1 liter is equivalent to 1 quart.

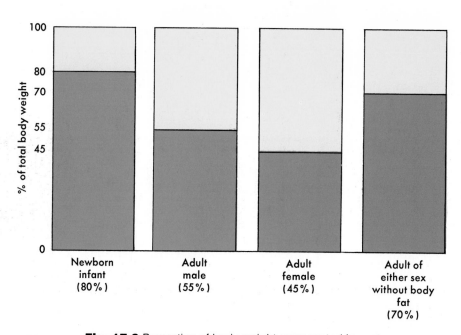

Fig. 17-2 Proportion of body weight represented by water.

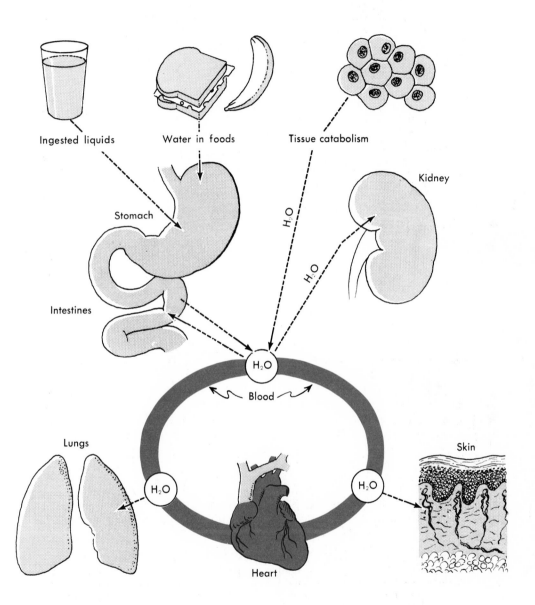

Fig. 17-3 The three sources of water entry into the body and four avenues of exit.

Table 17-1 Electrolyte composition of blood plasma (mEq*)	
Cations	**Anions**
142 mEq Na$^+$	102 mEq Cl$^-$
4 mEq K$^+$	26 mEq HCO$_3^-$
5 mEq Ca^{++}	17 mEq protein$^-$
2 mEq Mg^{++}	6 mEq other
	2 mEq HPO$_4^-$
153 mEq/L plasma	153 mEq/L plasma

*The milliequivalent (mEq) is a unit of measurement that indicates how reactive a particular electrolyte is in body-fluids.

intake, the less the urine volume. How changes in urine volume come about was discussed on p. 211. This would be a good time to review these paragraphs.

Several factors act as mechanisms for controlling plasma, interstitial fluid, and intracellular fluid volumes. We shall limit our discussion to naming only three of these factors, stating their effects on fluid volumes, and giving some specific examples of these effects. (You can find fuller discussions of the mechanisms involved in the supplementary readings for this chapter given on p. 259.) Three of the main factors that influence extracellular and intracellular fluid volumes are:

1. The concentration of electrolytes in the extracellular fluid
2. The capillary blood pressure
3. The concentration of proteins in blood

Importance of electrolytes in body fluids

The bonds that hold the molecules of certain organic substances such as glucose together are such that they do not permit the compound to break up or *dissociate* in solution. Such compounds are called *nonelectrolytes*. Compounds such as ordinary table salt or sodium chloride (NaCl) that have molecular bonds that permit them to break up or dissociate in solution into separate particles (Na$^+$ and Cl$^-$) are known as *electrolytes*.* The dissociated particles of an electrolyte are called *ions* and carry an electrical charge. Positively charged particles such as Na$^+$ are called *cations* and negatively charged particles such as Cl$^-$ are called *anions*. A variety of anions and cations serve important nutrient or regulatory roles in the body. Important cations include sodium (Na$^+$), calcium (Ca^{++}), potassium (K$^+$), and magnesium (Mg^{++}). Important anions include chloride (Cl$^-$), bicarbonate (HCO$_3^-$), phosphate (HPO$_4^-$), and many proteins. Table 17-1 shows that although blood

*Electrolytes are compounds that dissociate in solution to yield positively charged particles (cations) and negatively charged particles (anions).

Fig. 17-4 Sodium-containing internal secretions. The total volume of these secretions may reach 8,000 or more milliliters in a 24-hour period.

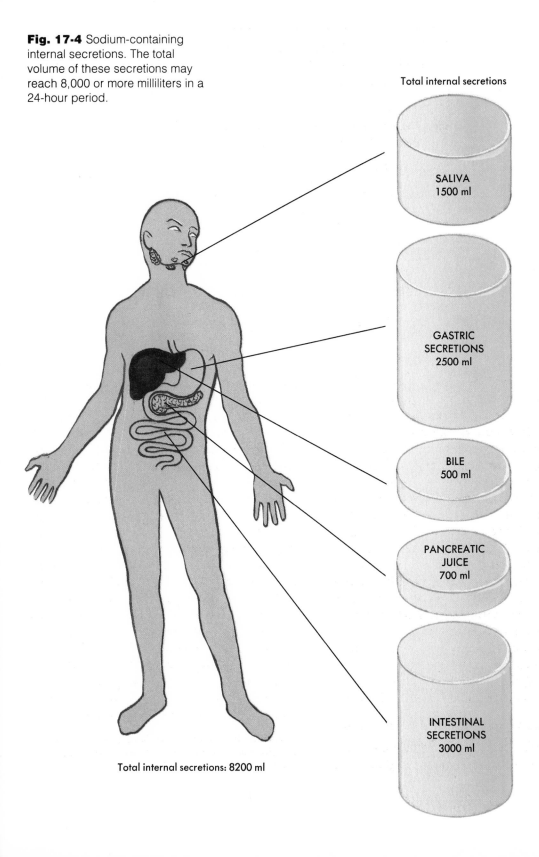

Total internal secretions

SALIVA
1500 ml

GASTRIC
SECRETIONS
2500 ml

BILE
500 ml

PANCREATIC
JUICE
700 ml

INTESTINAL
SECRETIONS
3000 ml

Total internal secretions: 8200 ml

plasma contains a number of important electrolytes, by far the most abundant one is sodium chloride (ordinary table salt, Na^+Cl^-). To remember how extracellular fluid electrolyte concentration affects fluid volumes, remember this one short sentence: where sodium goes, water soon follows. If, for example, the concentration of sodium in blood increases, the volume of blood soon increases. Conversely, if blood sodium concentration decreases, blood volume soon decreases.

Although wide variations are possible, the average daily diet contains about 100 mEq of sodium. In a healthy individual sodium excretion from the body by the kidney is about the same as intake. The kidney acts as the chief regulator of sodium levels in body fluids. It is important to know that many electrolytes such as sodium not only pass into and out of the body but also move back and forth between a number of body fluids during each 24-hour period. Fig. 17-4 shows the large volumes of sodium containing internal secretions that are produced each day. During a 24-hour period, over 8 liters of fluid containing 1,000 to 1,300 mEq of sodium are poured into the digestive system. This sodium, along with most of that contained in the diet, is almost completely reabsorbed. Very little sodium is lost in the feces. Precise regulation and control of sodium levels are required for survival.

Capillary blood pressure and blood proteins

Capillary blood pressure is a "water-pushing" force. It tends to push fluid out of the blood that is in capillaries into the interstitial fluid. Therefore if capillary blood pressure increases, more fluid is pushed—filtered—out of blood into the interstitial fluid. The effect of an increase in capillary blood pressure, then, is to transfer fluid from blood to interstitial fluid. In turn this fluid shift, as it is called, changes both blood and interstitial fluid volumes. It decreases blood volume by increasing interstitial fluid volume. If, on the other hand, capillary blood pressure decreases, less fluid filters out of blood into interstitial fluid.

Water continually moves in both directions through the membranous walls of capillaries. How much water moves out of capillary blood into interstitial fluid depends largely on capillary blood pressure, a water-pushing force. How much water moves in the opposite direction, that is, into blood from interstitial fluid, depends largely on the concentration of proteins present in blood plasma. Plasma proteins act as a water-pulling or water-holding force. They tend to hold water in the blood and to pull it into the blood from interstitial fluid. If, for example, the concentration of proteins in blood decreases appreciably—as it does in some abnormal conditions—less water moves into blood from interstitial fluid. As a result, blood volume decreases and interstitial fluid volume increases. Of the three main body fluids, interstitial fluid volume varies the most. Plasma volume usually fluctuates only slightly and briefly. If a pronounced change in its volume occurs, adequate circulation cannot be maintained.

Fluid imbalances

Fluid imbalances are common ailments. They take several forms and stem from a variety of causes, but they all share a common characteristic—that of abnormally low or abnormally high volumes of one or more body fluids. *Dehydration* is the fluid imbalance seen most often. In this potentially dangerous condition, interstitial fluid volume decreases first, but eventually, if treatment has not been given, intracellular fluid and plasma volumes also decrease below normal levels. Either too small a fluid intake or too large a fluid output causes dehydration. *Overhydration* can also occur but is much less common than dehydration. The grave danger of giving intravenous fluids too rapidly or in too large amounts is overhydration, which can put too heavy a burden on the heart.

Outline summary

Body fluids

A Major locations—outside of cells in interstitial spaces (IF) and in blood vessels (plasma) and inside cells (ICF)

B Percentage of body weight and volumes, assuming that weight is 120 pounds:

ECF (extracellular fluid)		
Plasma	4%	2.4 liters
IF (interstitial fluid)	16%	9.6 liters
ICF (intracellular fluid)	40%	24 liters
Total body fluid	60%	36 liters

C Variation in total body water related to:

1 Total body weight—the more a person weighs, the more water his body contains

2 Fat content of body—the more fat present, the less total water content per unit of weight (Fat is almost water free.)

3 Sex—proportion of body weight represented by water is about 10% less in females (45% than in males (55%)

4 Age—in newborn infant water may account for 80% of total body weight—see Fig. 17-2

Mechanisms that maintain fluid balance

A Fluid output, mainly urine volume, adjusts to fluid intake; ADH from posterior pituitary gland acts to increase kidney tubule reabsorption of water from tubular urine into blood, thereby tending to increase ECF (and total body fluid) by decreasing urine volume

B ECF electrolyte concentration (mainly Na^+ concentration) influences ECF volume; an increase in ECF Na^+ tends to increase ECF volume by increasing osmosis (movement of water) out of ICF and by increasing ADH se-

cretion, which decreases urine volume, and this, in turn, tends to increase ECF volume

C Capillary blood pressure tends to push water out of blood, into IF; blood protein concentration tends to pull water into blood from IF; hence these two forces regulate plasma and IF volume under usual conditions

D Importance of electrolytes in body fluids

1 Nonelectrolytes—organic substances that do not break up or dissociate when placed in solution; for example, glucose

2 Electrolytes—compounds that break up or dissociate in solution into separate particles called ions, for example, ordinary table salt or sodium chloride

$$NaCl \rightarrow Na^+ + Cl^-$$

3 Ions—the dissociated particles of an electrolyte that carry an electrical charge, for example, sodium ion (Na^+)

4 Cations—positively charged ions, for example, potassium (K^+), sodium (Na^+)

5 Anions—negatively charged particles (ions), for example, chloride (Cl^-), bicarbonate (HCO_3^-)

6 Electrolyte composition of blood plasma—see Table 17-1

7 Sodium—most abundant and important plasma cation

  a Normal plasma level—142 mEq/L

  b Average daily intake (diet)—100 mEq

  c Chief method of regulation—kidney

  d Sodium containing internal secretions—see Fig. 17-4

E Capillary blood pressure and blood proteins

Fluid imbalances

A Dehydration—total volume of body fluids smaller than normal; IF volume shrinks first, and then if treatment is not given, ICF volume decreases, and finally, plasma volume; dehydration occurs whenever fluid output exceeds fluid intake for an extended period of time; various factors may cause this; for example, diarrhea, gastrointestinal drainage or suction, or hemorrhage may increase fluid output above intake level; no liquid or food intake may decrease fluid intake below output level

B Overhydration—total volume of body fluids

larger than normal; IF volume expands first, and then if treatment is not given, ICF volume increases, and finally, plasma volume increases above normal; overhydration occurs whenever fluid intake exceeds fluid output; various factors may cause this; for example, giving excessive amounts of intravenous fluids or giving them too rapidly may increase intake above output; kidney failure or prolonged hypoventilation (depressed respirations) may decrease output below intake level

Review questions

1 Suppose a person who had never heard the term "fluid balance" were to ask you what it meant. How would you explain it briefly and simply?
2 Approximately what percentage of a slender adult's body weight consists of water?
3 The volume of blood plasma in a normal-sized adult weighs approximately what percentage of body weight?
4 The proportion of body weight represented by water is about 10% higher in males than in females. Why?
5 ICF makes up approximately what percentage of adult body weight?
6 Interstitial fluid makes up approximately what percentage of adult body weight?
7 To maintain fluid balance, does output usually change to match intake or does intake usually adjust to output?
8 Explain in words and by a diagram how ADH functions to maintain fluid balance.
9 Define the following terms:
  anion   ion
  cation   nonelectrolyte
  electrolyte
10 List the important anions and cations present in blood plasma.
11 Use the phrase, "where sodium goes, water soon follows," to explain how extracellular fluid electrolyte concentration affects fluid volumes.

12 List the sodium-containing internal secretions.

13 Explain by words or diagram how capillary blood pressure and blood protein concentration function to maintain fluid balance.

14 Suppose that an individual has suffered a hemorrhage and that, as a result, his capillary blood pressure has decreased below normal. What change would occur in blood and interstitial fluid volumes as a result of this decrease in capillary blood pressure?

15 Suppose that an individual has a type of kidney disease that allows plasma proteins to be lost in the urine and that, as a result, his plasma protein concentration decreases. How would this tend to change blood and interstitial fluid volumes?

16 If an individual becomes dehydrated, which fluid volume decreases first and which decreases last?

chapter 18

Acid-base balance

One of the requirements for healthy survival is that the body maintain, or quickly restore, the acid-base balance of its fluids. Maintaining acid-base balance means keeping the concentration of hydrogen ions in body fluids relatively constant. This is a matter of vital importance. If hydrogen ion concentration veers away from normal even slightly, cellular chemical reactions cannot take place normally and survival is thereby threatened.

pH of body fluids

Water and all water solutions contain both hydrogen ions (H^+) and hydroxyl ions (OH^-). The term "pH" followed by a number indicates a fluid's hydrogen-ion concentration. More specifically, pH 7.0 means an equal concentration of hydrogen ions and hydroxyl ions. Therefore, pH 7.0 also means that a fluid is neutral in reaction, that is, neither acid nor alkaline. The pH of water, for example, is 7.0. A pH higher than 7.0 indicates an alkaline solution, that is, one with a lower concentration of hydrogen ions than hydroxyl ions. The more alkaline a solution, the higher its pH. A pH lower than 7.0 indicates an acid solution, that is, one with a higher hydrogen-ion concentration than hydroxyl-ion concentration. The higher the hydrogen-ion concentration the lower the pH and the more acid a solution is. Normally, the pH of arterial blood is about 7.45 and the pH of venous blood is about 7.35. By applying information given in the last few sentences, you can deduce the answers to the following questions. Is arterial blood slightly acid or slightly alkaline? Is venous blood slightly acid or slightly alkaline? Which is a more accurate statement—venous blood is more acid than arterial blood or venous blood is less alkaline than arterial blood?*

Mechanisms that control pH of body fluids

The body has three mechanisms for regulating the pH of its fluids. They are the buffer mechanism, the respiratory mechanism, and the urinary mechanism. Together they constitute the complex pH homeostatic mechanism—the machinery that normally keeps blood slightly alkaline with a pH that stays remarkably constant. Its usual limits are very narrow, about 7.35 to 7.45.

The slight increase in acidity of venous blood (pH 7.35) compared to arterial blood (pH 7.45) results primarily from carbon dioxide entering venous blood as a waste product of cellular metabolism. As carbon dioxide enters the blood, some of it combines with water and is converted into car-

*Both arterial and venous blood are slightly alkaline because both have a pH slightly higher than 7.0. Venous blood, however, is less alkaline than arterial blood because venous blood's pH of about 7.35 is slightly lower than arterial blood's pH of 7.45.

bonic acid by *carbonic anhydrase*, an enzyme found in red blood cells:

$$CO_2 + H_2O \xrightarrow{\text{carbonic anhydrase}} H_2CO_3$$

The lungs remove the equivalent of over 30 liters of carbonic acid each day from the venous blood by elimination of carbon dioxide. This almost unbelievable quantity of acid is so well buffered that a liter of venous blood contains only about $1/100,000,000$ gram more hydrogen ions that does 1 liter of arterial blood. What incredible constancy! The pH homeostatic mechanism does indeed control effectively—astonishingly so.

Buffers

Buffers are substances that prevent a sharp change in the pH of a fluid when an acid or base is added to it. Strong acids and bases, if added to blood, would "dissociate" almost completely and release large quantities of hydrogen (H^+) or hydroxyl (OH^-) ions. The result would be drastic changes in blood pH. Survival itself depends on protecting the body from such drastic pH changes. More acids than bases are usually added to body fluids. This is because catabolism, a process that goes on continually in every cell of the body, produces acids that enter blood as it flows through tissue capillaries. Almost immediately, one of the salts present in blood—a buffer, that is—reacts with these relatively strong acids to change them to weaker acids. The weaker acids decrease blood pH only slightly, whereas the stronger acids formed by catabolism would have decreased it greatly if they were not buffered.

Buffers consist of two kinds of substances and are therefore often called "buffer pairs." One of the main blood "buffer pairs" is ordinary baking soda (sodium bicarbonate or $NaHCO_3$) and carbonic acid (H_2CO_3).

Let us consider, as a specific example of buffer action, how the sodium bicarbonate ($NaHCO_3$)–carbonic acid (H_2CO_3) system works in the presence of a strong acid or base.

Addition of a strong acid, such as hydrochloric acid (HC1), to the sodium bicarbonate–carbonic acid buffer system would initiate the reaction shown in Fig. 18-1. Note how this reaction between HC1 and sodium bicarbonate ($NaHCO_3$) applies the principle of buffering. As a result of the buffering action of $NaHCO_3$, the weak acid, $H{\cdot}HCO_3$, replaces the very strong acid, HCl, and therefore the hydrogen ion concentration of the blood increases much less than it would have if HCl were not buffered.

If, on the other hand, a strong base such as sodium hydroxide (NaOH) were added to the same buffer system, the reaction shown in Fig. 18-2 would take place. The hydrogen ion of carbonic acid ($H{\cdot}HCO_3$), the weak acid of the buffer pair, combines with the hydroxyl ion (OH^-) of the strong base sodium hydroxide (NaOH) to form water. Note what this accomplishes. It decreases the number of hydroxyl ions added to the solution, and this in turn prevents the drastic rise in pH that would occur in the absence of buffering.

Fig. 18-1 shows how a buffer system works in the presence of a strong acid. Although useful in demonstrating the principles of buffer action, HC1 or similar strong acids are never introduced directly into body fluids under normal circumstances. Instead, the sodium bicarbonate buffer system is most often called on to buffer a number of weaker acids produced during catabolism. Lactic acid is a good example. As a weak acid, it does not "dissociate" as completely as HCl. Incomplete dissociation of lactic acid results in fewer hydrogen ions being added to the blood and a less drastic lowering of blood pH than would occur if HCl were added in an equal amount. In the absence of buffering, however, lactic acid build-up will result in significant hydrogen ion accumulation over a period of time. The resulting decrease in pH can produce a serious acidosis. Ordinary baking soda (sodium bicarbonate or $NaHCO_3$) is one of the main buffers of the normally occurring "fixed" acids in blood. Lactic acid is one of the most abundant of the "fixed" acids, that is, acids that are not volatile and do not break down to form a gas. The following equation shows the compounds formed by

$$HCl + NaHCO_3 \rightarrow NaCl + H \cdot HCO_3$$

$$\underset{\text{(many)}}{H^+ + Cl^-} \qquad \underset{\text{(few)}}{H^+ + HCO_3^-}$$

Fig. 18-1 Buffering of acid HCl by sodium bicarbonate. As a result of buffer action, the strong acid (HCl) is replaced by a weaker acid ($H{\cdot}HCO_3$). Note that HCl as a strong acid "dissociates" almost completely and releases more hydrogen ions (H^+) than carbonic acid. Buffering decreases the number of hydrogen ions in the system.

$$NaOH + H{\cdot}HCO_3 \rightarrow NaHCO_3 + HOH$$

$$\underset{\text{(many)}}{Na^+ + OH^-} \qquad \underset{\text{(very few)}}{H^+ + OH^-}$$

Fig. 18-2 Buffering of base NaOH by carbonic acid. As a result of buffer action, the strong base (NaOH) is replaced by sodium bicarbonate and water. As a strong base, NaOH "dissociates" almost completely and releases large quantities of hydroxyl (OH^-) ions. Dissociation of water is minimal. Buffering decreases the number of hydroxyl ions in the system.

$$H \cdot lactate + NaHCO_3 \longrightarrow Na \cdot lactate + H \cdot HCO_3$$

$$H^+ + lactate^-$$
(few)

$$H^+ + HCO_3^-$$
(fewer)

Fig. 18-3 Lactic acid (H·lactate) and other nonvolatile or "fixed" acids are buffered by sodium bicarbonate in the blood. Carbonic acid (H·HCO$_3$ or H$_2$CO$_3$, a weaker acid than lactic acid) replaces lactic acid. As a result, fewer hydrogen ions are added to blood than would be if lactic acid were not buffered.

the buffering of lactic acid (a "fixed" acid), produced by normal body catabolism:

$$H \cdot lactate + NaHCO_3 \longrightarrow H \cdot HCO_3 + Na \cdot lactate$$

Lactic acid   Carbonic acid

Fig. 18-3 indicates the changes in blood that result from the buffering of fixed or nonvolatile acids in the tissue capillaries.

1. The amount of carbonic acid in blood increases slightly—because the nonvolatile acid (lactic acid in this case) is converted to volatile carbonic acid.
2. The amount of bicarbonate in blood (mainly sodium bicarbonate) decreases because bicarbonate ions become part of the newly formed carbonic acid. Normal arterial blood with a pH of 7.4 contains twenty times more sodium bicarbonate than carbonic acid. If this ratio decreases, blood pH decreases below 7.4.
3. The hydrogen ion concentration of blood increases slightly. Carbonic acid adds hydrogen ions to blood, but it adds fewer of them than lactic acid would have because carbonic acid is a weaker acid than lactic acid. In other words the buffering mechanisms does not totally prevent blood hydrogen-ion concentration from increasing. It simply minimizes the increase.
4. Blood pH decreases slightly because of

the small increase in blood hydrogen ion concentration.

Carbonic acid is the most abundant acid in body fluids because it is formed by the buffering of fixed acids and also because carbon dioxide forms carbonic acid by combining with water. Large amounts of carbon dioxide, an end product of catabolism, continually pour into tissue capillary blood from cells. Much of the carbonic acid formed in blood diffuses into red blood cells where it is buffered by the potassium salt of hemoglobin. Carbonic acid is a volatile acid; it breaks down to form the gas, carbon dioxide, and water. This takes place in blood as it moves through the lung capillaries. Read the next paragraphs to find out how this affects blood pH.

Respiratory mechanism of pH control

Respirations play a vital part in controlling pH. With every expiration, carbon dioxide and water leave the body in the expired air. The carbon dioxide has diffused out of the venous blood as it moves through the lung capillaries. Less carbon dioxide therefore remains in the arterial blood leaving the lung capillaries, so less carbon dioxide is available for combining with water to form carbonic acid. Hence arterial blood contains less carbonic acid and fewer hydrogen ions and has a higher pH (7.45) than does venous blood (pH 7.35).

Let us consider now how a change in respirations can change blood pH. Suppose you were to pinch your nose shut and hold your breath for a full minute or a little longer. Obviously, no carbon dioxide would leave your body by way of the expired air during that time and the blood's carbon dioxide content would necessarily increase. This would increase the amount of carbonic acid and the hydrogen-ion concentration of blood, which in turn would decrease blood pH. Here then are two useful facts to remember. Anything that causes an appreciable decrease in respirations will in time produce acidosis. Conversely, anything that causes an excessive increase in respirations will in time produce alkalosis (see p. 254).

Urinary mechanism of pH control

Most people know that the kidneys are vital organs and that life soon ebbs away if they stop functioning. One reason is that the kidneys are the body's most effective regulators of blood pH. They can eliminate much larger amounts of acid than can the lungs and, if it should become necessary, the kidneys can also excrete excess base. The lungs cannot. In short, the kidneys are the body's last and best defense against wide variations in blood pH. If they fail, homeostasis of pH—acid-base balance—fails.

Because more acids than bases usually enter blood, more acids than bases are usually excreted by the kidneys. In other words, most of the time the kidneys acidify urine; that is, they excrete enough acid to give urine an acid pH—frequently as low as 4.8. (How does this compare with normal blood pH?) The distal tubules of the kidneys rid the blood of excess acid and at the same time conserve the base present in it by the two mechanisms illustrated by Figs. 18-4 and 18-5. To understand these figures, you need to know only basic chemistry. If you have this knowledge, look at Fig. 18-4 and find the carbon dioxide leaving the blood (as it flows through a kidney capillary) and entering one of the cells that helps form the wall of a distal kidney tubule. Note that in this cell the carbon dioxide combines with water to form carbonic acid (H_2CO_3). This occurs rapidly because the cell contains car-

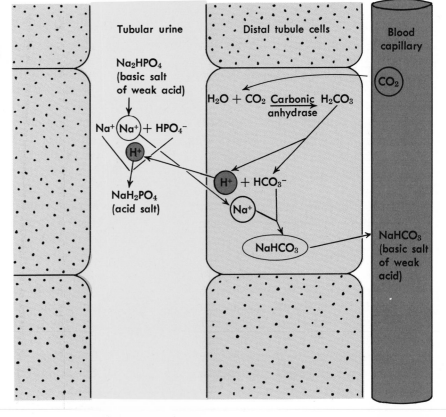

Fig. 18-4 Acidification of urine and conservation of base by distal renal tubule excretion of H ions (see text on this page).

bonic anhydrase, an enzyme that accelerates this reaction. As soon as carbonic acid forms, some of it dissociates to yield hydrogen ions and bicarbonate ions. Note what happens to these ions. Hydrogen ions diffuse out of the tubule cell into the urine trickling down the tubule. Here it replaces one of the sodium ions in a salt (Na_2HPO_4) to form another salt (NaH_2PO_4), which leaves the body in the urine. Notice next that the Na^+ displaced from Na_2HPO_4 by the hydrogen ion moves out of the tubular urine into a tubular cell. Here it combines with a bicarbonate (HCO_3^-) ion to form sodium bicarbonate, which then is reabsorbed into the blood. What this complex of reactions has accomplished is to add hydrogen ions to the urine—that is, acidify it—and to conserve sodium bicarbonate by reabsorbing it into the blood.

Fig. 18-5 illustrates another method of acidifying urine, as explained in the legend.

pH imbalances

Acidosis and alkalosis are the two kinds of pH or acid-base imbalance. In acidosis the blood pH falls as hydrogen ion concentration increases. Only rarely does it fall as low as 7.0 (neutrality) and almost never does it become even slightly acid, as death usually intervenes before the pH drops this much. In alkalosis, which develops less often than acidosis, the blood pH is higher than normal.

From a clinical standpoint disturbances in acid-base balance can be considered dependent on the relative quantities (ratio) of carbonic acid and sodium bicarbonate pres-

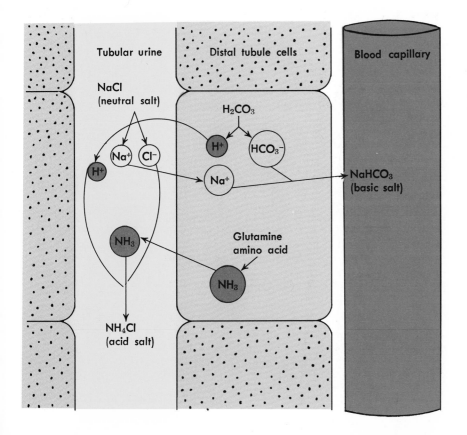

Fig. 18-5 Acidification of urine by tubule excretion of ammonia (NH_3). An amino acid (glutamine) leaves blood, enters a tubule cell, and is deaminized to form ammonia, which is excreted into urine. In exchange the tubule cell reabsorbs a basic salt (mainly $NaHCO_3$) into blood from urine.

ent in the blood. Components of this important "buffer pair" must be maintained at the proper ratio (twenty times more sodium bicarbonate than carbonic acid) if acid-base balance is to remain normal.* It is fortunate that the body can regulate both chemicals in the sodium bicarbonate–carbonic acid buffer system. Blood levels of sodium bicarbonate can be regulated by the kidneys and carbonic acid levels can be regulated by the respiratory system (lungs).

Metabolic and respiratory disturbances

Two types of disturbances, metabolic and respiratory, can alter the proper ratio of these components. Metabolic disturbances affect the bicarbonate element of the buffer pair, and respiratory disturbances affect the carbonic acid element, as follows:
1. Metabolic disturbances
   a. Metabolic acidosis (bicarbonate deficit)
   b. Metabolic alkalosis (bicarbonate excess)
2. Respiratory disturbances
   a. Respiratory acidosis (carbonic acid excess)
   b. Respiratory alkalosis (carbonic acid deficit)

The *ratio* of sodium bicarbonate to carbonic acid levels in the blood is the key to acid base balance. If the normal ratio (20 to 1 sodium bicarbonate to carbonic acid) can be maintained, the acid-base balance and pH will remain normal despite changes in the absolute amounts of either component of the buffer pair in the blood.

*Actually, in a state of acid-base balance a liter of plasma contains 27 mEq of sodium bicarbonate ($NaHCO_3$) or ordinary baking soda and 1.3 mEq of carbonic acid (H_2CO_3):

$$\frac{27 \text{ mEq } NaHCO_3}{1.3 \text{ mEq } H_2CO_3} = \frac{20}{1} = pH\ 7.4$$

As a clinical example, in a person suffering from untreated diabetes, abnormally large amounts of acids enter the blood. The normal 20 to 1 ratio of sodium bicarbonate to carbonic acid will be altered as the sodium bicarbonate component of the "buffer pair" reacts with the acids. Blood levels of sodium bicarbonate decrease rapidly in these patients. The result will be a lower ratio of sodium bicarbonate to carbonic acid (perhaps 10 to 1) and lower blood pH. The condition is called *uncompensated metabolic acidosis*. The body will attempt to correct or *compensate* for the acidosis by altering the *ratio* of $NaHCO_3$ to H_2CO_3. Acidosis in a diabetic patient is often accompanied by rapid breathing or hyperventilation. This compensatory action of the respiratory system results in a "blow-off" of carbon dioxide. Decreased blood levels of CO_2 result in lower carbonic acid levels. A new compensated ratio of sodium bicarbonate to carbonic acid (perhaps 10 to 0.5) may result. In such individuals the blood pH would return to normal or near normal levels. The condition is called *compensated metabolic acidosis*.

Outline summary
pH of body fluids
A Definition of pH—a number that indicates the hydrogen-ion concentration of a fluid; pH 7.0 indicates neutrality, pH higher than 7.0 indicates alkalinity, and pH less than 7.0 indicates acidity
B Normal arterial blood pH—about 7.45
C Normal venous blood pH—about 7.35

Mechanisms that control pH of body fluids
A Buffers
  1 Definition—buffers are substances that prevent a sharp change in the pH of a fluid when an acid or base is added to it—see Figs. 18-1 and 18-2
  2 Nonvolatile acids are buffered mainly by sodium bicarbonate ($NaHCO_3$)

3 Changes in blood produced by buffering of nonvolatile acids in the tissue capillaries
  a Amount of carbonic acid (H_2CO_3) in blood increases slightly
  b Amount of sodium bicarbonate in blood decreases; ratio of amount of $NaHCO_3$ to the amount of H_2CO_3 does not normally change; normal ratio is 20:1
  c Hydrogen ion concentration of blood increases slightly
  d Blood pH decreases slightly below arterial level
B Respiratory mechanism of pH control—respirations remove some of the carbon dioxide from blood; as blood flows through lung capillaries, the amount of carbonic acid in blood is decreased and thereby its hydrogen ion concentration is decreased, and this in turn increases blood pH from its venous to its arterial level
C Urinary mechanism of pH control—the body's most effective regulator of blood pH; kidneys usually acidify urine by the distal tubules excreting hydrogen ions and ammonia (NH) into the urine from blood in exchange for sodium bicarbonate being reabsorbed into the blood

pH imbalances

A Acidosis and alkalosis are the two kinds of pH or acid-base imbalance
B Disturbances in acid-base balance depend on relative quantities of sodium bicarbonate and carbonic acid in the blood
C Body can regulate both components of the $NaHCO_3$–H_2CO_3 buffer system
  1 Blood levels of $NaHCO_2$ regulated by kidneys
  2 Carbonic acid levels regulated by lungs
D Two basic types of pH disturbances, metabolic and respiratory, can alter the normal 20 to 1 ratio of $NaHCO_3$ to H_2CO_3 in blood
  1 Metabolic disturbances affect the sodium bicarbonate levels in blood
  2 Respiratory disturbances affect the carbonic acid levels in blood
E Types of pH or acid-base imbalances
  1 Metabolic disturbances
    a Metabolic acidosis (bicarbonate deficit)
    b Metabolic alkalosis (bicarbonate excess)
  2 Respiratory disturbances
    a Respiratory acidosis (carbonic acid excess)
    b Respiratory alkalosis (carbonic acid deficit)
F Clinical example
  1 Uncompensated metabolic acidosis
  2 Compensated metabolic acidosis

Review questions

1 Explain briefly what the term "pH" means.
2 What is the normal range for blood pH?
3 What is a typical normal pH for venous blood? for arterial blood?
4 When the body is in acid-base balance, arterial blood contains how many times more base bicarbonate (mainly $NaHCO_3$) than carbonic acid? In other words, what is the normal base bicarbonate/carbonic acid ratio in blood?
5 What function do buffers serve?
6 Explain how respirations affect blood pH.
7 Explain how the kidneys function to maintain normal blood pH.
8 How does prolonged hyperventilation (abnormally increased respirations) affect blood pH?
9 Which is the most important for maintaining acid-base balance—buffers, respirations, or kidney functioning?
10 Briefly, how does compensated acidosis differ from uncompensated acidosis? Define each of these terms briefly.
11 List the types of pH or acid-base imbalances.

Common medical abbreviations, prefixes, and suffixes

Abbreviations

āā of each
a.c. before meals
ad lib. as much as desired
alb. albumin
A.M. before noon
amt. amount
ante before
aq. water
Av. average
Ba barium
b.i.d. twice a day
b.m. bowel movement
B.M.R. basal metabolic rate
B.P. blood pressure
B.R.P. bathroom privileges
BUN blood urea nitrogen
c̄ with
C.B.C. complete blood count
C.C.U. coronary care unit
C.H.F. congestive heart failure
C.N.S. central nervous system
Co cobalt
C.V.A. cerebrovascular accident, stroke
D. & C. dilation and curettage
d/c discontinue
D.O.A. dead on arrival
Dx diagnosis
ECG electrocardiogram
E.D.C. expected date of confinement
EEG electroencephalogram
E.E.N.T. ear, eye, nose, throat
EKG electrocardiogram
E.R. Emergency room
F.U.O. fever of undetermined origin
G.I. gastrointestinal
G.P. general practitioner

GU genitourinary
h. hour
HCT hematocrit
Hgb hemoglobin
H₂O water
h.s. at bedtime
I.C.U. intensive care unit
K.U.B. kidney, ureter, and bladder
M.I. myocardial infarction
non rep. do not repeat
N.P.O. nothing by mouth
O.R. operating room
p.c. after meals
per by
P.H. past history
P.I. previous illness
P.M. after noon
p.c. after meals
p.r.n. as needed
q. every
q.d. every day
q.h. every hour
q.i.d. four times a day
q.n.s. quantity not sufficient
q.o.d. every other day
q.s. quantity sufficient
R.B.C. red blood cell
℞ prescription
s̄ without
sp. gr. specific gravity
s̄s̄. one half
stat. at once, immediately
T. & A. tonsillectomy and adenoidectomy
T.b. tuberculosis
t.i.d. three times a day
T.P.R. temperature, pulse, respiration
T.U.R. transurethral resection
W.B.C. white blood cell

Prefixes

a- without
ab- away from
ad- to, toward
adeno- glandular
amphi- on both sides
an- without
ante- before, forward
anti- against
bi- two, double, twice
circum- around, about
contra- opposite, against
de- away from, from
di- double
dia- across, through
dis- separate from, apart
dys- difficult
e- out, away
ecto- outside
en- in
endo- in, inside
epi- on
eu- well
ex- from, out of, away from
exo- outside
extra- outside, beyond; in addition
hemi- half
hyper- over, excessive, above
hypo- under, deficient
infra- underneath, below
inter- between, among
intra- within, on the side
intro- into, within
iso- equal, like
para- beside
peri- around, beyond
post- after, behind
pre- before, in front of
pro- before, in front of
re- again
retro- backward, back
semi- half
sub- under, beneath

super- above, over
supra- above, on upper side
syn- with, together
trans- across, beyond
ultra- excessive

Suffixes

-algia pain, painful
-asis condition
-blast young cell
-cele swelling
-centesis puncture for aspiration
-cide killer
-cyte cell
-ectomy cut out
-emia blood
-genesis production, development
-itis inflammation
-kinin motion, action
-logy study of
-megaly enlargement
-odynia pain
-oid resembling
-oma tumor
-osis condition
-opathy disease
-penia abnormal reduction
-pexy fixation
-phagia eating, swallowing
-phasia speaking condition
-phobia fear
-plasty plastic surgery
-plegia paralysis
-poiesis formation
-ptosis downward displacement
-rhaphy suture
-scope instrument for examination
-scopy examination
-stomy creation of an opening
-tomy incision
-uria urine

Suggested supplementary readings

1 An introduction to the structure and function of the body

Anthony, C. P., and Thibodeau, G. A.: Textbook of anatomy and physiology, ed. 10, St. Louis, 1979, The C. V. Mosby Co., pp. 17-23.

2 Cells and tissues

Anthony, C. P., and Thibodeau, G. A.: Basic concepts in anatomy and physiology: a programmed presentation, ed. 4, St. Louis, 1980, The C. V. Mosby Co., pp. 1-21.

Anthony, C. P., and Thibodeau, G. A.: Textbook of anatomy and physiology, ed. 10, St. Louis, 1979, The C. V. Mosby Co., pp. 28-67.

3 Organs and systems

Anthony, C. P., and Thibodeau, G. A.: Textbook of anatomy and physiology, ed. 10, St. Louis, 1979, The C. V. Mosby Co., pp. 80-147.

Roberts, S. L.: Skin assessment for color and temperature, Am. J. Nurs. **75:**610-613, 1975.

4 The skeletal system

Sonstegard, D. A., Mathews, L. S., and Kaufer, H.: The surgical replacement of the human knee joint, Sci. Am. **238:**44-51, 1978.

5 The muscular system

Anthony, C. P., and Thibodeau, G. A.: Textbook of anatomy and physiology, ed. 10, St. Louis, 1979, The C. V. Mosby Co. pp. 148-198.

Cohen, C.: The protein switch of muscle contraction, Sci. Am. **233:**36-45, 1975.

Hoyle, G.: How is muscle turned on and off? Sci. Am. **222:**84-93, 1970.

Murray, J. M., and Weber, A.: The cooperative action of muscle proteins, Sci. Am. **230:**59-71, 1974.

6 The nervous system

Anthony, C. P., and Thibodeau, G. A.: Basic concepts in anatomy and physiology: a programmed presentation, ed. 4, St. Louis, 1980, The C. V. Mosby Co., pp. 22-54.

Anthony, C. P., and Thibodeau, G. A.: Textbook of anatomy and physiology, ed. 10, St. Louis, 1979, The C. V. Mosby Co., pp. 220-275; 290-317.

7 The autonomic nervous system

Anthony, C. P., and Thibodeau, G. A.: Basic concepts in anatomy and physiology: a programmed presentation, ed. 4, St. Louis, 1980, The C. V. Mosby Co., pp. 55-63.

Anthony, C. P., and Thibodeau, G. A.: Textbook of anatomy and physiology, ed. 10., St. Louis, 1979, The C. V. Mosby Co., pp. 276-289.

8 The endocrine system

Anthony, C. P., and Thibodeau, G. A.: Basic concepts in anatomy and physiology: a programmed presentation, ed. 4, St. Louis, 1980, The C. V. Mosby Co., pp. 64-83.

Anthony, C. P., and Thibodeau, G. A.: Textbook of anatomy and physiology, ed. 10, St. Louis, 1979, The C. V. Mosby Co. pp. 318-347.

Krueger, J. A., and Ray, J. C.: Endocrine problems in nursing: a physiologic approach, St. Louis, 1976, The C. V. Mosby Co.

McEwen, B. S.: Interactions between hormones and nervous tissue, Sci. Am. **235:**48-67, 1976.

9 Blood

Anthony, C. P., and Thibodeau, G. A.: Basic concepts in anatomy and physiology: a programmed presentation, ed. 4, St. Louis, 1980, The C. V. Mosby Co., pp. 84-98.

Anthony, C. P., and Thibodeau, G. A.: Textbook of anatomy and physiology, ed. 10, St. Louis, 1979, The C. V. Mosby Co., pp. 350-367.

Bennett, B., et al.: The normal coagulation mechanism, Med. Clin. North Am. **59:**95, 1972.

Erskine, A. G., and Socha, W.: The principles and practice of blood grouping, ed. 2, St. Louis, 1978, The C. V. Mosby Co.

10 The cardiovascular and lymphatic systems

Abrahams, P., and Webb, P.: Clinical anatomy of practical procedures, Philadelphia, 1975, J. B. Lippincott Co., pp. 19-42.

Anthony, C. P., and Thibodeau, G. A.: Basic concepts in anatomy and physiology: a programmed presentation, ed. 4, St. Louis, 1980, The C. V. Mosby Co., pp. 99-123.

Anthony, C. P., and Thibodeau, G. A.: Textbook of anatomy and physiology, ed. 10, St. Louis, 1979, The C. V. Mosby Co., pp. 368-438.

Berne, R. M., and Levy, M. N.: Cardiovascular physiology, ed. 3, St. Louis, 1977, The C. V. Mosby Co.

Garrison, D.: Cardiovascular angiography for technologists, St. Louis, 1979, The C. V. Mosby Co.

Mayerson, H.: The lymphatic circulation, Sci. Am. **208:**80-93, 1963.

Phibbs, B.: The human heart: a guide to heart disease, ed. 3, St. Louis, 1975, The C. V. Mosby Co.

11 The immune system

Anthony, C. P., and Thibodeau, G. A.: Textbook of anatomy and physiology, ed. 10., St. Louis, 1979, The C. V. Mosby Co., pp. 650-663.

Donley, D. L.: Nursing the patient who is immunosuppressed, Am. J. Nurs. **76:**1619-1625, 1976.

Glasser, R. J.: The body is the hero, New York, 1976, Random House, Inc.

Jerne, N. K.: The immune system, Sci. Am. **229:**52-60, 1973.

Mayer, M. M.: The complement system, Sci. Am. **229:**54-66, 1973.

12 The digestive system

Anthony, C. P., and Thibodeau, G. T.: Textbook of anatomy and physiology, ed. 10, St. Louis, 1979, The C. V. Mosby Co., pp. 470-501; 506-533.

13 The respiratory system

Abrahams, P., and Webb, P.: Clinical anatomy of practical procedures, Philadelphia, 1975, J. B. Lippincott Co., pp. 47-69.

Anthony, C. P., and Thibodeau, G. A.: Basic concepts in anatomy and physiology: a programmed presentation, ed. 4, St. Louis, 1980, The C. V. Mosby Co., pp. 124-142.

Anthony, C. P., and Thibodeau, G. A.: Textbook of anatomy and physiology, ed. 10, St. Louis, 1979, The C. V. Mosby Co., pp. 440-469.

Comroe, J. H., Jr.: The lung, Sci. Am. **214:**57-66, 1966.

Egan, D. F.: Fundamentals of respiratory therapy, ed. 3, St. Louis, 1977, The C. V. Mosby Co.

Glover, D. W., and Glover, M. M.: Respiratory therapy: basics for nursing and the allied health professions, St. Louis, 1978, The C. V. Mosby Co.

Ruppel, G.: Manual of pulmonary function testing, St. Louis, 1979, The C. V. Mosby Co.

14 The urinary system

Anthony, C. P., and Thibodeau, G. A.: Basic concepts in anatomy and physiology: a programmed presentation, ed. 4, St. Louis, 1980, The C. V. Mosby Co., pp. 159-167.

Anthony, C. P., and Thibodeau, G. A.: Textbook of anatomy and physiology, ed. 10, St. Louis, 1979, The C. V. Mosby Co., pp. 538-561.

15 The male reproductive system

Anthony, C. P., and Thibodeau, G. A.: Basic concepts in anatomy and physiology: a programmed presentation, ed. 4, St. Louis, 1980, The C. V. Mosby Co., pp. 178-186.

Anthony, C. P., and Thibodeau, G. A.: Textbook of anatomy and physiology, ed. 10, St. Louis, 1979, The C. V. Mosby Co., pp. 606-619.

Epel, D.: The program of fertilization, Sci. Am. **237:**128-139, 1977.

16 The female reproductive system

Anthony, C. P., and Thibodeau, G. A.: Basic concepts in anatomy and physiology: a programmed presentation, ed. 4, St. Louis, 1980, The C. V. Mosby Co., pp. 187-200.

Anthony, C. P., and Thibodeau, G. A.: Textbook of anatomy and physiology, ed. 10, St. Louis, 1979, The C. V. Mosby Co., pp. 620-647.

Beer, A. E., and Billingham, R. E.: The embryo as a transplant, Sci. Am. **230:**36-51, 1974.

17 Fluid and electrolyte balance

Anthony, C. P., and Thibodeau, G. A.: Basic concepts in anatomy and physiology: a programmed presentation, ed. 4, St. Louis, 1980, The C. V. Mosby Co., pp. 201-208.

Anthony, C. P., and Thibodeau, G. A.: Textbook of anatomy and physiology, ed. 10, St. Louis, 1979, The C. V. Mosby Co., pp. 564-581.

Burke, S. R.: The composition and function of body fluids, ed. 2, St. Louis, 1976, The C. V. Mosby Co.

Gamble, J. L.: Chemical anatomy, physiology, and pathology of extracellular fluid, ed. 6, Cambridge, Mass., 1958, Harvard University Press.

Weldy, N. J.: Body fluids and electrolytes, ed. 2, St. Louis, 1976, The C. V. Mosby Co.

18 Acid-base balance

Anthony, C. P., and Thibodeau, G. A.: Basic concepts in anatomy and physiology: a programmed presentation, ed. 4, St. Louis, 1980, The C. V. Mosby Co., pp. 209-214.

Anthony, C. P., and Thibodeau, G. A.: Textbook of anatomy and physiology, ed. 10, St. Louis, 1979, The C. V. Mosby Co., pp. 582-594.

Davenport, H. W.: The ABC of acid-base chemistry, ed. 6, Chicago, 1973, University of Chicago Press.

Glossary

abdomen (ab-do′men; ab′do-men) body area between the diaphragm and pelvis.

abduct (ab-dukt′) to move away from the midline; opposite of adduct.

absorption (ab-sorp′shun) passage of a substance through a membrane (for example, skin or mucosa) into blood.

acetabulum (as″e-tab′u-lum) socket in the hipbone (os coxa or innominate bone) into which the head of the femur fits.

Achilles tendon (ah-kil′ēz ten′dun) tendon inserted on calcaneus; so called because of the Greek myth that Achilles' mother held him by the heels when she dipped him in the river Styx, thereby making him invulnerable except in this area.

acidosis (as″i-do′sis) condition in which there is an excessive proportion of acid in the blood.

acromegaly (ak″ro-meg′ah-le) condition caused by hypersecretion of growth hormone after puberty.

acromion (ah-kro′me-on) bony projection of the scapula; forms point of the shoulder.

adduct (ah-dukt′) to move toward the midline; opposite of abduct.

adenohypophysis (ad″ĕ-no-hi-pof′i-sis) anterior pituitary gland.

adenoid (ad′ĕ-noid) literally, glandlike; adenoids, or pharyngeal tonsils, are paired lymphoid structures in the nasopharynx.

adolescence (ad″o-les′ens) period between puberty and adulthood.

adrenergic fibers (ad″ren-er′jik fi′bers) axons whose terminals release norepinephrine and epinephrine.

afferent neuron (af′er-ent nu′ron) transmitting impulses to the central nervous system.

albuminuria (al″byu-mĭ-nu′re-ah) albumin in the urine.

aldosterone (al-dos′te-rōn) hormone secreted by adrenal cortex.

alkalosis (al″kah-lo′sis) condition in which there is an excessive proportion of alkali in the blood; opposite of acidosis.

alveolus (al-ve′o-lus) literally a small cavity; alveoli of lungs are microscopic saclike dilatations of terminal bronchioles.

amenorrhea (ah-men″o-re′ah) absence of menses.

amino acid (a-mee′no as′id) organic compound having an NH_3 and a COOH group in its molecule; has both acid and basic properties; amino acids are the structural units from which proteins are built.

amphiarthrosis (am″fe-ar-thro′sis) slightly movable joint.

anabolism (ah-nab′o-lizm) synthesis by cells of complex compounds (for example, protoplasm and hormones) from simpler compounds (amino acids, simple sugars, fats, and minerals); opposite of catabolism, the other phase of metabolism.

anaphase (an′a-fāz) stage of mitosis; duplicate chromosomes move to poles of dividing cell.

anastomosis (ah-nas″to-mo′sis) connection between vessels; the circle of Willis, for example, is an anastomosis of certain cerebral arteries.

androgen (an′dro-jen) male sex hormone.

anemia (ah-ne′me-ah) deficient number of red blood cells or deficient hemoglobin.

anesthesia (anes-the′ze-ah) loss of sensation.

aneurysm (an′u-rizm) blood-filled saclike dilatation of the wall of an artery.

angina (an-ji′nah) any disease characterized by spasmodic suffocative attacks; for example, angina pectoris and paroxysmal thoracic pain with feeling of suffocation.

angstrom (ang′strum) unit 0.1 μu ($^1/_{10,000,000,000}$ of a meter or about 1/250,000,000 of an inch).

anions (an′-i-unz) negatively charged particles.

ankylosis (ang″kĭ-lo′sis) abnormal immobility of a joint.

anorexia (an″o-rek′se-ah) loss of appetite.

anoxia (an-ok′se-ah) deficient oxygen supply to tissues.

antagonistic muscles (an-tag′o-nis-tik mus′el) those having opposing actions; for example, muscles that flex the upper arm are antagonists to muscles that extend it.

anterior (an-te′re-or) front or ventral; opposite of posterior or dorsal.

antibody, immune body (an′tĭ-bod″e, ĭ-myun′ bod′e) substance produced by the body that destroys or inactivates a specific substance (antigen) that has entered the body; for example, diphtheria antitoxin is the antibody against diphtheria toxin.

antigen (an′tĭ-jen) substance that, when introduced into the body, causes formation of antibodies against it.

antrum (an′trum) cavity; for example, the antrum of Highmore, the space in each maxillary bone, or the maxillary sinus.

anus (a′nus) distal end or outlet of the rectum.

apex (a′peks) pointed end of a conical structure.

aphasia (ah-fa′ze-ah) loss of a language faculty, such as the ability to use words or understand them.

apnea (ap-ne′ah) temporary cessation of breathing.

aponeurosis (ap″o-nu-ro′sis) flat sheet of white fibrous tissue that serves as a muscle attachment.

arachnoid (ah-rak′noid) delicate, weblike middle membrane of the meninges.

areola (ah-re′o-lah) small space; the pigmented ring around the nipple.

arteriole (ar-te′re-ōl) small branch of an artery.

arteriosclerosis (ar-ter′i-o-scle-ro′sis) hardening of the arteries.

artery (ar′ter-e) vessel carrying blood away from the heart.

arthrosis (ar-thro′sis) joint or articulation.

articulation (ar-tik-u-la′shun) joint.

ascites (ah-si′tēz) accumulation of serous fluid in the abdominal cavity.

asphyxia (as-fik′se-ah) loss of consciousness caused by deficient oxygen supply.

aspirate (as′pŭ-rāt) to remove by suction.

ataxia (ah-tak′se-ah) loss of power of muscle coordination.

atherosclerosis (ath-e-ro″-sclero-′sis) hardening of arteries, lipid deposits in lining coat.

atrium (a′tre-um) chamber or cavity; for example, atrium of each side of the heart.

atrophy (at′ro-fe) wasting away of tissue; decrease in size of a part.

auricle (aw′re-kl) part of the ear attached to the side of the head; earlike appendage of each atrium of heart.

autonomic (aw″to-nom′ik) self-governing; independent.

axilla (ak-sil′ah) armpit.

axon (ak′son) nerve cell process that transmits impulses away from the cell body.

B

Bartholin (bar′to-lin) seventeenth century Danish anatomist.

basophil (ba′so-fil) white blood cell that stains readily with basic dyes.

biceps (bi′seps) a muscle having two heads.

bilirubin (bil″e-roo′bin) red pigment in the bile.

biliverdin (bil″e-ver′din) green pigment in the bile.

Bowman (Bo′man) nineteenth century English physician.

brachial (brak′e-al) pertaining to the arm.

bronchiectasis (brong″ke-ek′tah-sis) dilatation of the bronchi.

bronchiole (brong′ke-ōl) small branch of a bronchus.

bronchus (brong′kus) one of the two branches of the trachea.

buccal (buk′al) pertaining to the cheek.

buffer (buf′er) compound that combines with an acid or with a base to form a weaker acid or base, thereby lessening the change in hydrogen-ion concentration that would occur without the buffer.

buffy coat (buff′e kot) layer of white cells located between plasma and packed red cells in a hematocrit tube.

bursa (bur′sah) fluid-containing sac or pouch lined with synovial membrane.

bursitis (bur-si′tis) inflammation of a bursa.

buttock (but′ok) prominence over the gluteal muscles.

c

calculus (kal′kyu-lus) stone; formed in various parts of the body; may consist of different substances.

calorie (kal′o-re) heat unit; a large calorie is the amount of heat needed to raise the temperature of 1 kilogram of water 1 degree Celsius.

calyx (ka′liks) cup-shaped division of the renal pelvis.

capillary (kap′i-la″re) microscopic blood vessel; capillaries connect arterioles with venules; also, microscopic lymphatic vessels.

carbaminohemoglobin (kar-bam″ĭ-no-he″mo-glo′bin) compound formed by union of carbon dioxide with hemoglobin.

carbohydrate (kar″bo-hi′drāt) organic compounds containing carbon, hydrogen, and oxygen in certain specific proportions; for example, sugars, starches, and cellulose.

carboxyhemoglobin (kar-bok″se-he″mo-glo′bin) compound formed by union of carbon monoxide with hemoglobin.

carcinoma (kar″sĭ-no′mah) cancer, a malignant tumor.

cardiopulmonary resuscitation (kar″de-o-pul′mo-ner-e re-sus″i-ta′shun) (CPR) combined cardiac (heart) massage and artificial respiration.

caries (ka′re-ēz) decay of teeth or of bone.

carotid (kah-rot′id) from Greek word *karos*, meaning "deep sleep"; carotid arteries of the neck so called because pressure on them may produce unconsciousness.

carpal (kar′pal) pertaining to the wrist.

casein (ka′se-in) protein in milk.

cast (kast) mold; for example, formed in renal tubules.

castration (kas-tra′shun) removal of testes or ovaries.

catabolism (kah-tab′o-lism) breakdown of food compounds or protoplasm into simpler compounds; opposite of anabolism, the other phase of metabolism.

catalyst (kat′ah-list) substance that accelerates the rate of a chemical reaction.

cataract (kat′ah-rakt) opacity of the lens of the eye.

catecholamines (kat″e-kol-am′inz) norepinephrine and epinephrine.

cations (cat′i-unz) positively charged particles.

cecum (se′kum) blind pouch; the pouch at the proximal end of the large intestine.

celiac (se′le-ak) pertaining to the abdomen.

cellulose (sel′yu-lōs) polysaccharide, the main plant carbohydrate.

centimeter (sen′tĭ-me″ter) 1/100 of a meter, about ²/₅ of an inch.

centrioles (sen′trĭ-ōlz) two dots seen with a light microscope in the centrosphere of a cell; active during mitosis.

cerumen (sĕ-roo′men) earwax.

cervicitis (ser″vi-si′tis) inflammation of the cervix.

cervix (ser′viks) neck; any neckline structure.

chiasm (ki′azm) crossing; specifically, a crossing of the optic nerves; also **chiasma**

cholecystectomy (ko″le-sis-tek′to-me) removal of the gallbladder.

cholesterol (ko-les′ter-ol) organic alcohol present in bile, blood, and various tissues.

cholinergic fibers (ko″lin-er′jik fi′bers) axons whose terminals release acetycholine.

cholinesterase (ko″lin-es′ter-ās) enzyme; catalyzes breakdown of acetylcholine.

chromatin (kro′mah-tin) deep-staining substance in the nucleus of cells; divides into chromosomes during mitosis.

chromosome (kro′mo-sōm) one of the segments into which chromatin divides during mitosis; involved in transmitting hereditary characteristics.

chyle (kīl) milky fluid; the fat-containing lymph in the lymphatics of the intestine.

chyme (kīm) partially digested food mixture leaving the stomach.

cilia (sil″e-ah) hairlike projections of protoplasm.

circumcision (ser″kum-sizh′un) surgical removal of the foreskin or prepuce.

circadian (ser″kah-de′an) daily.

cochlea (kok′le-ah) snail shell or structure of similar shape.

coenzyme (ko-en′zīm) nonprotein substance that activates an enzyme.

collagen (kol′ah-jen) principle organic constituent of connective tissue.

colloid (kol′oid) dissolved particles with diameters of 1 to 100 millimicrons (1 millimicron equals about 1/25,000,000 of an inch).

colostrum (ko-los′trum) first milk secreted after childbirth.

combining sites. (com-bin′ing sits) antigen-binding sites, antigen receptor regions on antibody molecule; shape of each combining site is complementary to shape of a specific antigen's epitopes.

complement (com′ple-ment) several inactive enzymes normally present in blood, which kill foreign cells by dissolving them.

concha (kong′kah) shell-shaped structure; for example, bony projections into the nasal cavity.

condyle (kon′dīl) rounded projection at the end of a bone.

congenital (kon-jen′ĭ-tal) present at birth.

contralateral (kon″trah-lat′er-al) on the opposite side.

coracoid (kor′ah-koid) like a raven's beak in form.

corium (ko′re-um) true skin or derma.

coronal (ko-ro′nal) like a crown.

coronary (kor′o-na-re) encircling; in the form of a crown.

corpus (kor′pus) body.

corpuscle (kor′pus″l) very small body or particle.

cortex (kor′teks) outer part of an internal organ, for example, of the cerebrum and of the kidneys.

cortisol (kor′ti-sol) the chief hormone secreted by the adrenal cortex; hydrocortisone; compound F.

costal (kos′tal) pertaining to the ribs.

crenation, plasmolysis (kre-na′shun, plazmol′ĭ-sis) shriveling of a cell caused by water withdrawal.

cretinism (kre′tin-izm) dwarfism caused by hypofunction of the thyroid gland.

cribriform (krib′rĭ-form) sievelike.

cricoid (kri′koid) ring-shaped; a cartilage of this shape in the larynx.

crystalloid (kris′tal-loid) dissolved particle less than 1 millimicron in diameter.

Cushing's syndrome (koosh′ingz sin′drom) condition caused by hypersecretion of glucocorticoids from the adrenal cortex.

cutaneous (kyu-ta′ne-us) pertaining to the skin.

cyanosis (si″ah-no′sis) bluish appearance of the skin caused by deficient oxygenation of the blood.

cytology (si-tol′o-je) study of cells.

cytoplasm (si′to-plasm″) the protoplasm of a cell exclusive of the nucleus.

D

deciduous (de-sid′yu-us) temporary; shedding at a certain stage of growth; for example, deciduous teeth.

decussation (de″kus-sa′shun) crossing over like an X.

defecation (def′e-ka′shun) elimination of waste matter from the intestines.

deglutition (de″gloo-tish′un) swallowing.

deltoid (del′toid) triangular; for example, deltoid muscle.

dendrite, dendron (den′dr-it, den′dron) branching or treelike; a nerve cell process that transmits impulses toward the cell body.

dens (denz) tooth.

dentate (den′tāt) having toothlike projections.

dentine, dentin (den′tēn; den′tin) main part of a tooth, under the enamel.

dentition (den-tish′un) teething; also, number, shape, and arrangement of the teeth.

dermis, corium (der′mis, ko′re-um) true skin.

dextrose (deks′trōs) glucose, a monosaccharide, the principal blood sugar.

diabetes insipidus (di″ah-be′tez in-sip-i-dus) condition characterized by a large urine volume caused by deficiency of antidiuretic hormone (ADH).

diabetes mellitus (di″ah-be′tez mel″li-tus) condition characterized by a high blood glucose level caused by a deficiency of insulin.

diaphragm (di′ah-fram) membrane or partition that separates one thing from another; the muscular partition between the thorax and abdomen; the midriff.

diaphysis (di-af′ĭ-sis) shaft of a long bone.

diarthrosis (di″ar-thro′sis) freely movable joint.

diastole (di-as′to-le) relaxation of the heart, interposed between its contractions; opposite of systole.

diastolic pressure (di″ah-stol′ik presh′ur) blood pressure in arteries during diastole (relaxation) of heart.

diencephalon (di″en-sef′ah-lon) "tween" brain; parts of the brain between the cerebral hemispheres and the mesencephalon or midbrain.

diffusion (dĭ-fyu′zhun) spreading; for example, scattering of dissolved particles.

digestion (di-jes′chun) conversion of food into assimilable compounds.

diplopia (dĭ-plo′pe-ah) double vision; seeing one object as two.

disaccharide (di-sak′ah-rīd) sugar formed by the union of two monosaccharides; contains twelve carbon atoms.

distal (dis′tal) toward the end of a structure; opposite of proximal.

diuresis (di″u-re′sis) increased urine production.

dorsal, posterior (dor′sal, pos-te′re-or) pertaining to the back; opposite of ventral; in man posterior is dorsal.

dropsy (drop′se) accumulation of serous fluid in a body cavity or in tissues; edema.

dura mater (du′rah ma′ter) literally strong or hard mother; outermost layer of the meninges.

dyspnea (disp′ne-ah) difficult or labored breathing.

dystrophy (dis′tro-fe) faulty nutrition.

E

ectopic (ek-top′ik) displaced; not in the normal place; for example, extrauterine pregnancy.

edema (e-de′mah) excessive fluid in tissues; dropsy.

effector (ef-fek′tor) responding organ; for example, voluntary and involuntary muscle, the heart, and glands.

efferent (ef′er-ent) carrying from, as neurons that transmit impulses from the central nervous system to the periphery; opposite of afferent.

electrocardiogram (e-lek″tro-kar′de-o-gram) graphic record of heart's action potentials.

electroencephalogram (e-lek″tro-en-sef′ah-lo-gram) graphic record of brain's action potentials.

electrolyte (e-lek′tro-l-it) substance that ionizes in solution, rendering the solution capable of conducting an electric current.

electron (e-lek′tron) minute, negatively charged particle.

elimination (e-lim″ĭ-na′shun) expulsion of wastes from the body.

embolism (em′bo-lizm) obstruction of a blood vessel by foreign matter carried in the bloodstream.

embolus (em′bo-lus) a blood clot or other substance (bubble of air) that is moving in the blood and may block a blood vessel.

embryo (em′bre-o) animal in early stages of intrauterine development; the human fetus the first 3 months after conception.

emesis (em′e-sis) vomiting.

emphysema (em″fĭ-se′mah) dilation of pulmonary alveoli.

empyema (em″pi-e′mah) pus in a cavity; for example, in the chest cavity.

encephalon (en-sef′ah-lon) brain.

endocrine (en′do-krin) secreting into the blood or tissue fluid rather than into a duct; opposite of exocrine.

endometritis (en″do-me-tri′tis) inflammation of the uterine lining or endometrium.

endoplasmic reticulum (en′do-plas′mic re-tic′u-lum) network tubules and vesicles in cytoplasm.

energy (en′er-je) capacity for doing work.

enteron (en′ter-on) intestine.

enzyme en'sīm) catalytic agent formed in living cells.

eosinophil, acidophil (e″o-sin′o-fil, asid′o-fil″) white blood cell readily stained by eosin.

epidermis (ep″ĭ-der′mis) "false" skin; outermost layer of the skin.

epinephrine (ep″ĭ-nef′rin) adrenaline; secretion of the adrenal medulla.

epiphyses (e-pif′ĭ-sēz) ends of a long bone.

erythrocyte (e-rith′ro-sīt) red blood cell.

ethmoid (eth′moid) sievelike.

eupnea (yoop-ne′ah) normal respiration.

Eustachio (ā-oo-stah′ke-o) sixteenth century Italian anatomist

eustachian canal (yu-sta′ke-an că-nal′) tube from inside of ear to throat to equalize air pressure.

exocrine (ek′so-krin) secreting into a duct; opposite of endocrine.

exophthalmos (ek″sof-thal′mos) abnormal protrusion of the eyes.

extrinsic (eks-trin′sik) coming from the outside; opposite of intrinsic.

F

fallopian tubes (fal-lo′pe-in toobs) pair of tubes that conduct ovum from ovary to uterus.

Fallopius (fal-lo′pe-us) sixteenth century Italian anatomist.

fascia (fash′e-ah) sheet of connective tissue.

fasciculus (fah-sik′yu-lus) little bundle.

fetus (fe′tus) unborn young, especially in the later stages; in human beings, from third month of intrauterine period until birth.

fiber (fi′ber) threadlike structure.

fibrin (fi′brin) insoluble protein in clotted blood.

fibrinogen (fi-brin′o-jen) soluble blood protein that is converted to insoluble fibrin during clotting.

fimbria (fim′bre-ah) fringe.

fissure (fish′ūr) groove.

flaccid (flak′sid) soft, limp.

follicle (fol′lĭ-k″l) small sac or gland.

fontanelles (fon″tah-nelz′) "soft spots" of the infant's head; unossified areas in the infant skull.

foramen (fo-ra′men), plural **foramina** (fo-ram′in-ah) small opening.

fossa (fos′sah) cavity or hollow.

fovea (fo′ve-ah) small pit or depression.

fundus (fun′dus) base of a hollow organ; for example, the part farthest from its outlet.

G

ganglion (gang′gle-on) cluster of nerve cell bodies outside the central nervous system.

gasserian (gas-se′re-an) named for Gasser, a sixteenth century Austrian surgeon; trigeminal ganglion.

gastric (gas′trik) pertaining to the stomach.

gene (jēn) part of the chromosome that transmits a given hereditary trait.

genitals (gen′i-t′lz) reproductive organs; genitalia.

gestation (jes-ta′shun) pregnancy.

gland (gland) secreting structure.

glomerulus (glo-mer′yu-lus) compact cluster; for example, of capillaries in the kidneys.

glossal (glos′al) of the tongue.

glucagon (gloo′kah-gon) hormone secreted by alpha cells of the islands of Langerhans.

glucocorticoids (gloo″ko-kor′tĭ-koidz) hormones that influence food metabolism; secreted by adrenal cortex.

gluconeogenesis (gloo″ko-ne″o-jen′e-sis) formulation of glucose or glycogen from protein or fat compounds.

glucose (gloo′kōs) monosaccharide or simple sugar; the principal blood sugar.

gluteal (gloo′te-al) of or near the buttocks.

glycerin, glycerol (glis′er-in, glis′er-ol) product of fat digestion.

glycogen (gli′ko-jen) polysaccharide; animal starch.

glycogenesis (gli″ko-jen′e-sis) formation of glycogen from glucose or from other monosaccharides, fructose or galactose.

glycogenolysis (gli″ko-jĕ-nol′ĭ-sis) hydrolysis of glycogen to glucose-6-phosphate or to glucose.

glyconeogenesis (gli″ko-ne″o-jen′e-sis) See **gluconeogenesis.**

goiter (goi′ter) enlargement of thyroid gland.

gonad (gon′ad) sex gland in which reproductive cells are formed.

graafian (graf′e-an) named for Graaf, a seventeenth century Dutch anatomist; pertaining to ovarian follicle.

gradient (gra′de-ent) a slope or difference between two levels; for example, blood pressure gradient—a difference between the blood pressure in two different vessels.

gustatory (gus'tah-to"re) pertaining to taste.

gyrus (ji"rus) convoluted ridge.

H

haversian (ha-ver'shan) named for Havers, English anatomist of late seventeenth century; pertaining to small blood vessels in bone.

heart block (hart blok) blockage of impulse conduction from atria to ventricles so that heart beats at a slower rate than normal.

hematocrit (he-mat'o-krit) volume percent of blood cells in whole blood.

hemiplegia (hem"e-ple'je-ah) paralysis of one side of the body.

hemoglobin (he"mo-glo'bin) iron-containing protein in red blood cells.

hemolysis (he-mol'ĭ-sis) destruction of red blood cells with escape of hemoglobin from them into surrounding medium.

hemopoiesis (he"mo-poi-e'sis) blood cell formation.

hemorrhage (hem'or-ij) bleeding.

hepar (he'par) liver.

heparin (hep'ah-rin) substances obtained from the liver; inhibits blood clotting.

heredity (he-red'ĭ-te) transmission of characteristics from a parent to a child.

hernia, "rupture" (her'ne-ah, rup'chur) protrusion of a loop of an organ through an abnormal opening.

hilus, hilum (hi'lus, hi'lum) depression where vessels enter an organ.

His (hiss) German anatomist of late nineteenth century.

histology (his-tol'o-je) science of minute structure of tissues.

homeostasis (ho"me-o-sta'sis) relative uniformity of the normal body's internal environment.

hormone (hor'mōn) substance secreted by an endocrine gland.

hyaline (hi'ah-lin) glasslike.

hydrocortisone (hi"dro-kor'tĭ-sōn) a hormone secreted by the adrenal cortex; cortisol; compound F.

hydrolysis (hi-drol'ĭ-sis) literally "splitting by water"; chemical reaction in which a com-

pound reacts with water to form simpler compounds.

hymen (hi'men) Greek for "membrane"; mucous membrane that may partially or entirely occlude the vaginal outlet.

hyoid (hi'oid) U-shaped; bone of this shape at the base of the tongue.

hyperemia (hi"per-e'me-ah) increased blood in a part of the body.

hyperkalemia (hi"per-kah-le'me-ah) higher than normal concentration of potassium in the blood.

hypernatremia (hi"per-na-tre'me-ah) higher than normal concentration of sodium in the blood.

hyperopia (hi"per-o'pe-ah) farsightedness.

hyperplasia (hi"per-pla'ze-ah) increase in the size of a part caused by an increase in the number of its cells.

hyperpnea (hi"perp-ne'ah) abnormally rapid breathing; panting.

hypertension (hi"per-ten'shun) abnormally high blood pressure.

hyperthermia (hi"per-ther'me-ah) fever; body temperature above 37° C.

hypertrophy (hi-per'tro-fe) increased size of a part caused by an increase in the size of its cells.

hypoglycemia (hi"po-gli-se'me'ah) abnormally low blood glucose level.

hypokalemia (hi"po-ka-le'me-ah) lower than normal concentration of potassium in the blood.

hyponatremia (hi"po-na-tre'me-ah) lower than normal concentration of sodium in the blood.

hypophysis (hi-pof'ĭ-sis) Greek for "outgrowth"; hence the pituitary gland, which grows out from the undersurface of the brain.

hypothermia (hi"po-ther'me-ah) subnormal body temperature below 37° C.

hypoxia (hi-pok'se-ah) oxygen deficiency.

I

incus (ing'kus) anvil; the middle ear bone that is shaped like an anvil.

inferior (in-fe're-or) lower; opposite of superior.

inguinal (ing'gwĭ-nal) of the groin.

inhalation (in"hă-la'shun) inspiration or breathing in; opposite of exhalation or expiration.

inhibition (in″hī-bish′un) checking or restraining of action.

innominate (in-nom′ĭ-nāt) not named, anonymous; for example, ossa coxae (hipbones), formerly known as innominate bones.

insulin (in′su-lin) hormone secreted by islands of Langerhans in the pancreas.

intercellular (in″ter-sel′yu-lar) between cells; interstitial.

internuncial (in″ter-nun′she-al) like a messenger between two parties; hence an internuncial neuron (or interneuron) is one that conducts impulses from one neuron to another.

interstitial (in″ter-stish′al) forming small spaces between things; intercelluar.

intrinsic (in-trin′sik) not dependent on externals; located within something; opposite of extrinsic.

involuntary (in-vol′un-ter″e) not willed; opposite of voluntary.

involution (in″vo-lu′shun) return of an organ to its normal size after enlargement; also retrograde or degenerative change.

ion (i′on) electrically charged atom or group of atoms.

ipsilateral (ip″sĭ-lat′er-al) on the same side; opposite of contralateral.

irritability (ir″ĭ-tah-bil′ĭ-te) excitability; ability to react to a stimulus.

ischemia (si-ke′me-ah) local anemia; temporary lack of blood supply to an area.

isometric (i″so-met′rik) type of muscle contraction in which muscle does not shorten.

isotonic (i″so-ton′ik) of the same tension or pressure.

K

ketones (ke′tōnz) acids (acetoacetic, beta-hydroxybutyric, and acetone) produced during fat catabolism.

ketosis (ke-to′sis) excess amount of ketone bodies in the blood.

kilogram (kil′o-gram) 1,000 grams; approximately 2.2 pounds.

kinesthesia (kin″es-the′ze-ah) "muscle sense"; that is, sense of position and movement of body parts.

L

labia (la′be-ah) lips.

lacrimal (lak′ri-mal) pertaining to tears.

lactation (lak-ta′shun) secretion of milk.

lactose (lak′tōs) milk sugar, a disaccharide.

lacuna (lah-kyu′nah) space or cavity; for example, lacunae in bone contain bone cells.

lamella (lah-mel′ah) thin layer, as of bone.

lateral (lat′er-al) of or toward the side; opposite of medial.

leukemia (lu-ke′me-ah) blood cancer characterized by an increase in white blood cells.

leukocyte (lu′ko-sīt) white blood cells.

leukocytosis (lu″ko-si-to′sis) abnormally high white blood cell numbers in the blood.

leukopenia (lu″ko-pe′ne′ah) abnormally low white blood cell numbers in the blood.

ligament (lig′ah-ment) bond or band connecting two objects; in anatomy a band of white fibrous tissue connecting bones.

lipid (lip′id) fats and fatlike compounds.

loin (loin) part of the back between the ribs and hipbones.

lumbar (lum′ber) of or near the loins.

lumen (loo′men), plural **lumina** (loo′mĭ-nah) passageway or space within a tubular structure

luteum (lu′te-um) golden yellow.

lymph (limf) watery fluid in the lymphatic vessels.

lymphocyte (lim′fo-sīt) one type of white blood cell.

lysosomes (li′so-sōmz) membranous organelles containing various enzymes that can dissolve most cellular compounds; hence called "digestive bags" or "suicide bags" of cells.

M

malleolus (mal-le′o-lus) small hammer; projections at the distal ends of the tibia and fibula.

malleus (mal′e-us) hammer; the tiny middle ear bone that is shaped like a hammer.

Malpighi (mal-pig′e) seventeenth century Italian anatomist.

maltose (mawl′tōs) disaccharide or "double" sugar.

mammary (mam′er-e) pertaining to the breast.

manometer (mah-nom′e-ter) instrument used for measuring the pressure of fluids.

manubrium (mah-nu′bre-um) handle; upper part of the sternum.

mastication (mas″ti-ka′shun) chewing.

matrix (ma′triks) ground substance in which cells are embedded.

meatus (me-a′tus) passageway.

medial (me′de-al) of or toward the middle; opposite of lateral.

mediastinum (me″de-as-ti′num) middle section of the thorax; that is, between the two lungs.

medulla (me-dul′lah) Latin for "marrow"; hence the inner portion of an organ in contrast to the outer portion, or cortex.

meiosis (mi-o′sis) nuclear division in which the number of chromosomes are reduced to half their original number before the cell divides in two.

membrane (mem′brăn) thin layer or sheet.

menarche (me-nar′ke) beginning of the menstrual function.

menopause (men′o-pawz) termination of menstrual cycles.

menstruation (men″stroo-a′shun) monthly discharge of blood from the uterus.

mesentery (mes′en-ter″e) fold of peritoneum that attaches the intestine to the posterior abdominal wall.

mesial (me′ze-al) situated in the middle; median.

metabolism (mĕ-tab′o-lizm) complex process by which food is utilized by a living organism.

metacarpus (met″ah-kar′pus) "beyond" the wrist; hence the part of the hand between the wrist and fingers.

metatarsus (met″ah-tar′sus) "beyond" the instep; hence the part of the foot between the tarsal bones and toes.

meter (me′ter) about 39.5 inches.

microglia (mi-krog′le-ah) one type of connective tissue found in the brain and cord.

micron (mi′kron) 1/1,000 millimeter; about 1/25,000 inch.

micturition (mik″tu-rish′un) urination, voiding.

millimeter (mil′ĭ-me-ter) 1/1,000 meter; about 1/25 inch.

mineralocorticoids (min″er-al-o-kor′tĭ-koidz) hormones that influence mineral salt metabolism; secreted by adrenal cortex; aldosterone is the chief mineralocorticoid.

mitochondria (mi″to-kon′dre-ah) threadlike structures.

mitosis (mi-to′sis) indirect cell division involving complex changes in the nucleus.

mitral (mi′tral) shaped like a miter.

monosaccharide (mon″o-sak′ah-rīd) simple sugar; for example, glucose.

motoneurons (mo″to-nu′ronz) cells that transmit nerve impulses away from the brain or spinal cord; also called motor, or efferent, neurons.

myasthenia gravis (mi″as-the′ne-ah gra′vis) disease marked by progressive muscular weakness.

myelin (mi′ĕ-lin) lipoid substance found in the myelin sheath around some nerve fibers.

myeloid (mi′e-loid) pertaining to bone marrow.

myocardial infarction (mi″o-kar′de-al infark′shun) death of cardiac muscle cells resulting from inadequate blood supply as in coronary thrombosis.

myocarditis (mi″o-kar-di′tis) inflammation of the myocardium (heart muscle).

myocardium (mi″o-kar′de-um) muscle of the heart.

myopia (mi-o′pe-ah) nearsightedness.

myxedema (mik″se-de′mah) condition caused by deficiency of thyroid hormone in adult.

N

nares (na′rēz) nostrils.

neurilemma, neurolemma (nu″rĭ-lem′mah) nerve sheath.

neuroglia (nu-rog′le-ah) fine-webbed supporting structure of nervous tissue.

neurohypophysis (nu″ro-hi-pof′ĭ-sis) posterior pituitary gland.

neuron (nu′ron) nerve cell, including its processes.

neutrophil (nu′tro-fil) white blood cell that stains readily with neutral dyes.

nucleus (nu′kle-us) spherical structure within a cell; a group of neuron cell bodies in the brain or cord.

O

occiput (ok′sĭ-put) back of the head.

olecranon (o-lek′rah-non) elbow.

olfactory (ol-fak'to-re) pertaining to the sense of smell.

ophthalmic (of-thal'mik) pertaining to the eyes.

organelle (or"gan-el') cell organ; one of the specialized parts of a single-celled organism (protozoon), serving for the performance of some individual function.

os (ahs) Latin for "mouth" (plural **ora**) and for "bone" (plural **ossa**).

osmosis (oz-mo'sis) movement of a fluid through a semipermeable membrane.

ossicle (os'sĭ-k'l) little bone.

oxidation (ok"si-da'shun) loss of hydrogen or electrons from a compound or element.

oxyhemoglobin (ok"se-he"mo-glo'bin) a compound formed by union of oxygen with hemoglobin.

P

palate (pal'ĭt) roof of the mouth.

palpebrae (pal'pe-bre) eyelids.

papilla (pah-pil'lah) small nipple-shaped elevation.

paralysis (pah-ral'ĭ-sis) loss of the power of motion or sensation, especially loss of voluntary motion.

parenchyma (par-eng'kĭ-mah) essential or functional tissue of an organ.

parietal (pah-ri'ĕ-tal) of the walls of an organ or cavity.

parotid (pah-rot'id) located near the ear.

parturition (par"tu-rish'un) act of giving birth.

patella (pah-tel'lah) small, shallow pan; the kneecap.

pectoral (pek'to-ral) pertaining to the chest or breast.

pelvis (pel'vis) basin or funnel-shaped structure.

pericardium (per"i-kar-de-um) membrane that surrounds the heart.

peripheral (pĕ-rif'er-al) pertaining to an outside surface.

peritoneum (per"i-to-ne'um) saclike membrane lining the abdominopelvic walls.

pH (pe'āch') hydrogen-ion concentration.

phagocytosis (fag"o-si-to'sis) ingestion and digestion of particles by a cell.

phalanges (fah-lan'jēz) finger or toe bones.

phrenic (fren'ik) pertaining to the diaphragm.

pia mater (pi'ah ma'ter) Latin for "gentle mother"; the vascular innermost covering (meninges) of the brain and cord.

pineal (pin'e-al) shaped like a pinecone.

piriformis (pir"i-for'mis) pear-shaped.

pisiform (pi'sĭ-form) pea-shaped.

plantar (plan'tar) pertaining to the sole of the foot.

plasma (plaz'mah) liquid part of the blood.

plasmolysis (plaz-mol'ĭ-sis) shrinking of a cell caused by water loss by osmosis.

pleurisy (ploor'i-se) inflammation of the pleura.

plexus (plek'sus) network.

pneumothorax (nu"mo-tho'raks) accumulation of air in the pleural space causing collapse of lung.

polycythemia (pol"e-si-the'me-ah) an excessive number of red blood cells.

polymorphonuclear (pol"e-mor"fo-nu'kle-ar) having many-shaped nuclei.

polysaccharide (pol"e-sak'ah-rīd) complex sugar.

pons (ponz) bridge.

popliteal (pop-lit'e-al) behind the knee.

posterior (pos-te're-or) located behind; opposite of anterior.

posture (pos'tur) position of the body.

presbyopia (pres"be-o'pe-ah) "oldsightedness"; farsightedness of old age.

pronate (pro'nāt) to turn palm downward.

proprioceptors (pro"pri-o-sep'tors) receptors located in the muscles, tendons, and joints.

prostaglandins (pros"tah-glan'dins) a group of naturally occuring fatty acids that affect many body functions.

protoplasm (pro'to-plazm) living substance.

proximal (prok'sĭ-mal) next or nearest; located nearest the center of the body or the point of attachment of a structure.

psoas (so'iss) pertaining to the loin, the part of the back between the ribs and hipbones.

psychosomatic (si"ko-so-mat'ik) pertaining to the influence of the mind, notably the emotions, on body functions.

puberty (pyu'ber-te) age at which the reproductive organs become functional.

Purkinje system (pur'kin'jē) specialized fibers in the heart that conduct cardiac impulses into the walls of the ventricles.

R

receptor (re-sep'tor) peripheral beginning of a sensory neuron's dendrite.

reflex (re'fleks) involuntary action.

refraction (re-frak'shun) bending of a ray of light as it passes from a medium of one density to one of a different density.

renal (re'nal) pertaining to the kidney.

residual volume (re-sid'u-al vol'um) air that remains in the lungs after the most forceful expiration.

reticular (re-tik'u-lar) netlike.

reticulum (re-tik'u-lum) a network.

ribosomes (ri'bo-sōmz) organelles in cytoplasm of cells; synthesize proteins, so nicknamed "protein factories."

rugae (roo'jee) wrinkles or folds.

S

sagittal (saj'ĭ-tal) like an arrow; longitudinal.

salpingitis (sal"pin-ji'tis) inflammation of the uterine (fallopian) tubes.

salpinx (sal'pinks) tube; oviduct.

sartorius (sar-to're-us) tailor; hence, one uses the thigh muscle to sit cross-legged like a tailor.

sciatic (si-at'ik) pertaining to the ischium.

sclera (skle'rah) from Greek for "hard"; white outer coat of eyeball.

scrotum (skro'tum) bag around testicles.

sebum (se'bum) Latin for "tallow"; secretion of sebaceous glands.

sella turcica (sel'ah tur'sikah) Turkish saddle; saddle-shaped depression in the sphenoid bone.

semen (se'men) Latin for "seed"; male reproductive fluid.

semilunar (sem"e-lu'nar) half-moon–shaped.

senescence (se-nes'ens) old age.

serratus (ser-ra'tus) saw-toothed.

serum (se'rum) any watery animal fluid; clear, yellowish liquid that separates from a clot of blood.

sesamoid (ses'ah-moid) shaped like a sesame seed.

sigmoid (sig'moid) C-shaped; S-shaped.

sinus (si'nus) cavity.

soleus (so'le-us) pertaining to a sole; a muscle in the leg shaped like the sole of a shoe.

somatic (so-mat'ik) of the body framework or walls, as distinguished from the viscera or internal organs.

spermatogenesis (sper"mah-to-jen'e-sis) production of sperm cells.

spermatozoa (sper"mah-to-zo'ah) sperm cells.

sphenoid (sfe'noid) wedge-shaped.

sphincter (sfingk'ter) ring-shaped muscle.

splanchnic (splank'nik) visceral.

squamous (skwa'mus) scalelike.

stapes (sta'pēz) stirrup; tiny stirrup-shaped bone in the middle ear.

stimulus (stim'u-lus) agent that causes a change in the activity of a structure.

stress (stress) according to Selye, physiological stress is a condition in the body produced by all kinds of injurious factors that he calls "stressors" and manifested by a syndrome (a group of symptoms that occur together).

stressor (stres'sor) any injurious factor that produces biological stress; for example, emotional trauma, infections, severe exercise.

striated (stri'āt-id) marked with parallel lines.

sudoriferous (su"dor-if'er-us) secreting sweat.

sulcus (sul'kus) furrow or groove.

superior (su-pe're-or) higher; opposite of inferior.

supinate (si'pĭ-nāt) to turn the palm of the hand upward; opposite of pronate.

surfactant (ser-fak'tant) substance that lines alveolar sacs to reduce surface tension.

Sylvius (sil've-us) seventeenth century Dutch anatomist; also a sixteenth century French anatomist.

symphysis (sim'fĭ-sis) Greek for "growing together"; place where two bones fuse.

synapse (sin'aps) joining; point of contact between adjacent neurons.

synergist (sin'er-jist) muscle that assists a prime mover.

synovia (sĭ-no've-ah) literally "with egg"; secretion of the synovial membrane; resembles egg white.

synthesis (sin'thĭ-sis) putting together of parts to form a more complex whole.

systole (sis'to-le) contraction of the heart muscle.

systolic pressure (sis'tol'ik) blood pressure in arteries during systole (contraction of heart).

T

talus (ta'lus) ankle; one of the bones of the ankle.

target organ cell (tahrget or'gan sel) organ (cell) acted on by a particular hormone and responding to it.

tarsus (tahr'sus) instep.

tendon (ten'dun) band or cord of fibrous connective tissue that attaches a muscle to a bone or other structure.

tetany (tet'ah-ne) continuous muscular contraction.

thorax (tho'raks) chest.

thrombin (throm'bin) protein important in blood clotting.

thrombosis (throm-bo'sis) formation of a clot in a blood vessel.

thrombus (throm'bus) stationary blood clot.

tibia (tib'e-ah) Latin for "shinbone."

tidal volume (tid'el vol'um) amount of air breathed in and out with each breath.

tonus (to'nus) continued, partial contraction of muscle.

tract (trakt) bundle of axons located within the central nervous system.

trauma (traw'mah) injury.

trochlear (trok'li-ar) pertaining to a pulley.

trophic (trof'ik) having to do with nutrition.

turbinate (tur'bĭ-nāt) shaped like a cone or like a scroll or spiral.

tympanum (tim'pah-num) drum.

U

umbilicus (um-bil'ĭ-kus) navel.

universal donor blood (yu-ne-versel do'nor blud) type O blood (contains neither A nor B antigens).

universal recipient blood (yu-ne-ver'sel re-sip'e-ent blud) type AB blood (contains neither anti-A nor anti-B antibodies).

utricle (yu'tre-k'l) little sac.

uvula (yu'vyu-lah) Latin for "little grape"; a projection hanging from the soft palate.

V

vagina (vah-ji'nah) sheath; internal tube from uterus to vulva.

vagus (va'gus) Latin for "wandering."

valve (valv) structure that permits flow of a fluid in one direction only.

vas (vass) vessel or duct.

vasectomy (vah-sek'to-me) surgical removal of a portion of the vas deferens to induce sterility.

vastus (vas'tus) wide; of great size.

vein (vān) vessel carrying blood toward the heart.

ventral (ven'tral) of or near the belly; in man, front or anterior; opposite of dorsal or posterior.

ventricle (ven'trĭ-k'l) small cavity.

vermiform (ver'mĭ-form) worm-shaped.

villus (vil'lus) hairlike projection.

viscera (vis'er-ah) internal organs (singular, viscus).

vital capacity (vi'tal kah-pas'i-te) largest amount of air that can be moved in and out of the lungs in one inspiration and expiration.

vomer (vo'mer) plowshare; unpaired flat bone of nasal septum.

vulva (vul'vah) external genitals of the female.

X

xiphoid (zif'oid) sword-shaped; pertaining to cartilaginous lower end of sternum.

Z

zygoma (zi-go'mah) "bar," "bolt"; zygomatic bone in cheek.

Index